D0382064

STEPHEN JAY GOULD and the politics of evolution

STEPHEN JAY GOULD and the politics of evolution

david f. prindle

Prometheus Books
59 John Glenn Drive
Amherst, New York 14228-2119

Published 2009 by Prometheus Books

Inquiries should be addressed to
Prometheus Books
59 John Glenn Drive
Amherst, New York 14228–2119
VOICE: 716–691–0133, ext. 210
FAX: 716–691–0137
WWW.PROMETHEUSBOOKS.COM

13 12 11 10 09   5 4 3 2 1

Library of Congress Cataloging-in-Publication Data

Prindle, David F. (David Forrest), 1948–
Stephen Jay Gould and the politics of evolution / by David F. Prindle.
p. cm.
Includes bibliographical references and index.
ISBN 978–1–59102–718–8 (hardcover : alk. paper)
1. Gould, Stephen Jay—Political and social views. 2. Evolution (Biology)—Political aspects. 3. Science—Political aspects. 4. Science—Philosophy. 5. Science and religion. I. Title.

QH31.G65P75 2009
576.8092—dc22

2009005251

Printed in the United States on acid-free paper

to my friends
mark lockhart and william alex rennie,
for decades of good conversation about science,
politics, and life.

contents

acknowledgments

As with any academic book, all of the chapters of this one went through many drafts and revisions. Sometimes I added, sometimes I subtracted, and sometimes I subtly modified. Occasionally I revised because of a new thought or a new reading, but just as often after a conversation with another scholar or a written critique from a reviewer. I cannot thank everyone by name who helped, because some were anonymous reviewers at academic presses. Those reviewers read the whole manuscript and gave me detailed critiques, and, if they chance upon this book in published form, I want them to know that I am grateful to them—even those who disliked my central thesis and therefore gave me a hard time. Hard times are useful.

But I can name some. Here are people who read all or part of individual chapters, and made various kinds of suggestions. I did not take all the suggestions, of course, but I considered each one at some length.

Chapter 1: Ellen Kogan, Jay Wilbur
Chapter 2: Richard Lewontin, Jay Wilbur
Chapter 3: Dan Bolnick, Niles Eldredge
Chapter 5: John Loehlin
Chapter 6: Thomas Gregg

Additionally, there is a group of people with whom I discussed many of the ideas contained herein. Some criticized, some suggested, but all helped. They are Roger Berkowitz, Walter Dean Burnham, Benjamin Gregg, Robert Hardgrave, Melvin Hinich, Donna Howell, Mark Lockhart, William Alex Rennie, and Robert Sprinkle.

I owe a particular debt to Professor Daniel Bolnick of the Section of Integrative Biology in the School of Biological Sciences at the University of Texas at Austin for permitting me to audit his graduate course on "Speciation" during the Spring 2006 semester. I guess, when I started, that I knew about as little about evolutionary biology as the average political scientist, so it took a good deal of forbearance by Dan and his graduate students to endure my habitual assertiveness in class. But they did endure it with good cheer, and I hope they all win Nobel prizes. The students were Mark Brinkman, Evan Economo, An Lee, Hernan Lopez-Fernandez, Jen Olori, Mary McGovern, Frank Stearns, and Roxi Steele.

introduction

One July day in 1969, while working my way through my undergraduate years as a rear chainman on a survey crew in Alaska, I inadvertently thrust my fist into a hornets' nest nestled within a patch of thick bushes. The next few minutes were among the most memorable of my life.

The years I have spent working on this book have been joyously entertaining, but they have also been memorable in a manner reminiscent of my experience in Alaska. Sometimes I have felt as if I had ignorantly stuck my hand into an academic hornets' nest.

I knew when I began the project that Stephen Jay Gould lived his professional life amidst scientific controversies, and that I would have to participate in those controversies in order to make sense of Gould's contribution to evolutionary biology. That knowledge did not intimidate me. Indeed, the existence of the controversies was partly what drew me to Gould and his ideas. I also knew that scholars can be aggressively territorial in their suspicion of people not trained in their specialties who presume to write books about them. I anticipated a good deal of skepticism from paleontologists, ethologists, and those in related fields concerning my qualifications, and I tried to forestall the accusation of poaching by reading professional literature, attending professional classes, participating in professional conventions, interviewing professional luminaries, and approaching professional controversies with as much humility as I could muster.

What I did not anticipate, however, was the hostility among some members of the fraternity of evolutionary biologists to my central thesis in this book, that Gould's mind worked along two tracks simultaneously, the scientific and the political. As an observer who is sensitive to political implication by temperament and training, it seemed glaringly obvious to me that Gould never penned a line that did not address, if only implicitly, both areas of human thought. It seemed too evident for contradiction that each of his essays on the history of life was also a meditation on larger political issues.

But my cavalier assumptions along these lines were mistaken, and called out the hornets. Evidently, many professional scientists perceive any effort to consider the political implications of scientific ideas to be an attack on the idea of scientific truth itself. To them, apparently, real science is simply an exercise in theorizing and considering evidence. Every attempt to consider the social implications or evaluate the personal interests or persuasive strategies of scientists is an effort to undermine the whole enterprise.

As a person who naturally thinks simultaneously in the political and scientific realms, I was caught up short when I encountered this attitude among natural scientists. When I first sent my manuscript around to academic presses (which shall remain unnamed), I was expecting specific criticisms and even general negativity from the internal reviewers—those are part of the scholarly life. But I was not prepared to read this sort of thing in a review:

> Science is not decided by vote but by evidence. It is not up to Gould to convince anyone of anything; it is up to scientists to pursue the evidence that leads them to support or reject hypotheses that have been proposed to explain the pattern and processes of the evolution of life. . . . It is a question of the standards of evidence and the methods accepted by the field that are of importance; nothing more. . . . What does it matter that a few people argue about political implications? They may be full of shit. . . . I think that his [Prindle's] main thesis, that Gould's science and his politics were inseparable, is overstated and unsupportable in many instances.

My project, in other words, was a waste of time. Either there are no political implications in science, or there are but they are irrelevant.

That particular reviewer, however, did not speak for all of them. Just as prevalent among the early internal reviews was this sort of statement:

> I don't think that Prindle is sufficiently critical of Gould regarding the degree to which Gould's political and social views biased his scientific views. . . . (On Gould's view that mass extinctions are fundamentally random) his political position was so prevalent in his thinking here that it caused him to abandon rational science. . . . Gould's affirmation of the ruling concept of chance did not lead him to conclusions about merit and equality. Rather . . . his social attitudes about merit and equality—not scientific inquiry—led him to his exaggerated views on the role of chance in evolution.

So, the first reviewer was upset with me because I ascribed too much political intention to Gould's science, and the second reviewer was upset with me because I did not ascribe enough political intention to Gould's science. At least the second reviewer did not reject my central thesis; in fact, he thought I did not run with it as far as I should have.

Since this is a book about a controversial figure, I'm not going to complain about coming in for my share of controversy in the way I treat him. But because the reviewers raise the issue, and because I deal with it in the text, I think I should clearly state a few of my own positions. I believe that there is such a thing as objective reality in nature, independent of the human mind. I believe that the scientific method is the best way to achieve successively more correct understandings of the way nature functions. I believe that evidence, reasoning, and ever-more-sophisticated methodology do, in the long run, help scientists to fashion progressively superior understandings of the natural world, including the part of the natural world that pertains to *Homo sapiens*. But I also believe (with Gould) that scientists always have personal and cultural preconceptions that affect the way they theorize about the world and interpret data. I believe that some scientific misunderstandings can endure for long periods of time because cultural conditionings warp the judgment of generations of scientists. I believe that most scientific discoveries have potential implications for economic and cultural relationships, that scientists are often aware of these potential implications, and that the awareness can influence their choice of projects, evaluation of each others' arguments, and willingness to accept or reject the results of a scientific test. Finally, I believe that some subject areas are inherently ambiguous, and that the attempt to investigate them with scientific methods is fated to always fail to achieve a definitive conclusion. I believe, in other words, that science is the best known method for getting close to the truth, but that it is always an imperfect enterprise.

As a noble but imperfect activity, science is bound to be fertile ground for a political scientist. And so, despite the hornets, my study of Stephen Jay Gould and the politics of evolution has been worthwhile; in fact, it has been fun.

Chapter 1

a charming style

I never met Stephen Jay Gould, but I wish I had. We could have had a pleasant conversation about dinosaurs. Like most boys with an intellectual bent, I grew up imagining the terrible lizards as pets, or as role-players in stories of my own invention. And as with many such boys, I eventually transferred my interests elsewhere, in my case from archaic monsters to contemporary ones, and went on to study human politics rather than natural history. Dinosaurs, however, remained for me part of the fascinating ephemera of science, one of those subjects that retained a magnetic glamour in the world outside my own profession.

In a similar manner, Gould made extinct reptiles a focus of his youthful enthusiasm. But unlike so many of the rest of us, he managed to achieve the status of an intellectual Peter Pan by contriving a way to make a living by never growing up. He went into a scientific discipline, paleontology, that specializes in the study of long-dead organisms. By thus turning his boyhood obsession into his adult profession, he succeeded in living out a myth of personal fulfillment. And by communicating his boyish enthusiasm to an audience of millions, he drew them into his private myth. In the process, he became perhaps America's best-known scientist.

The myth he lived and communicated, however, was not without self-consciousness and artifice. The famous Stephen Jay Gould, the "evolutionist laureate," the best-selling author, the media performer, the national treasure, did not simply stumble, boyishly, into the status of public intellectual. As an adult, he used the myth of the enthusiastic boy now grown up to further his personal and intellectual goals. And the manner of his creation of this charming myth bears examination. As with many other public personae, there was more to Stephen Jay Gould than met the eye.

Gould began to build the myth of the boyhood ambition fulfilled in *The Panda's Thumb,* his second collection of monthly essays from *Natural History* magazine, published in 1980. "When I was four I wanted to be a garbageman," he begins one of the essays, and goes on to describe his career epiphany:

> Then, when I was five, my father took me to see the *Tyrannosaurus* at the American Museum of Natural History. As we stood in front of the beast, a man sneezed; I gulped . . . but the great animal stood immobile in all its bony grandeur, and as we left, I announced that I would be a paleontologist when I grew up.[1]

Gould does not want us to forget this story, for he repeats it in two of his other books.[2] Further, his essays are full of efforts to communicate his own still-youthful passion for solving the puzzles of life's history. "The boy dinosaur enthusiast still dwells within me,"[3] he tells us in another of his collections. Sustaining the mood, he assures his readers that "science . . . is the greatest of human adventures,"[4] and that "the best scientists live a life of keen amusement."[5] For his part, he tells us that the "main motivation" for the composition strategy he adopted in another book was "simple joy."[6] Unlike many scientists, who like to portray their profession as consisting mainly of careful plodding, Gould does not hesitate to proclaim that science is fun!

But the key to understanding Gould's appeal, of course, is to realize that doing paleontology was not just fun for him; he made it fun for us. Gould was more than a scientist. He was also an eloquent, inspiring writer. He did not just educate his readers; he charmed them.

Although in his final book he made the self-deprecating comment that he had "literary pretensions," in fact his standing as a superb writer was ratified by the clutch of awards his books won over the years.[7] *The Panda's Thumb* (1980) garnered the National Book Award; *The Mismeasure of Man* (1981) won a National Book Critics Circle Award; *Hen's Teeth and Horse's Toes* (1983) captured the Phi Beta Kappa Book Award; and *Wonderful Life* (1989) was a finalist for the Pulitzer Prize. Beyond the level of literary prizes, his standing as a scientist who wrote books that were accessible and entertaining to nonscientists made him far more than a Harvard scholar. He became a fixture in American culture. He wrote the preface to one of Gary Larson's *Far Side* cartoon collections. He voiced his own character in an episode of the television series *The Simpsons.* He was interviewed for stories in *Harper's*, *Newsweek*, *Time*, and *The New York Times Magazine,* and was the subject of a profile in *People.* He was elected president of the American Association for the Advancement of Science. His 1989 book *Wonderful Life* became a best-seller in three countries. In 2001 the Library of Congress named Gould one of the country's "Living Legends," a group of people who each embody the "quintessentially American ideal of individual creativity, conviction, dedication, and exuberance."[8] When the National Center for Science Education created a project to advance the teaching of evolution in classrooms, they named it "Project Steve," after Gould.[9] Until his death at the early age of sixty in 2002, he truly was, as philosopher Daniel Dennett proclaimed him, "America's evolutionist laureate."[10]

But not all were charmed. While Gould was building a reputation as *the* evolutionary essayist among the public, he was creating an often intense opposition to his person and his ideas within the academic industry of evolutionary biology. Dennett himself, after acknowledging Gould's great fame, devoted four chapters in his own book on evolutionary theory to savaging the laureate's ideas.[11] Biologist John Alcock wrote an article in a professional journal in which he complained of "the little tricks and standard misrepresentations embedded in Gould's critiques."[12] Psychologist J. Philippe Rushton asserted in a review of the second edition of *The Mismeasure of Man* that it was "a political polemic, whose author engages in character assassination of long deceased scientists whose work he misrepresents despite published refutations, while studiously withholding from his readers 15 years of new research that contradicts every major scientific argument he puts forth."[13] It would be possible, in fact, to fill a chapter with denunciations and critiques of Gould's work by other scholars. There is a contradiction here that needs to be explored.

Part of Gould's interest as a scientist, and much of the opposition his publications generated, can be attributed to the fact that he was not just a scholar. He was also a concerned and participating citizen with strongly held political ideas. Moreover—and here I come to the reason why a political scientist would write a book about Gould—his scientific ideas were seamlessly wedded to his political positions, so that his methodological and philosophical stance always buttressed his political values and vice-versa. In the record of human thought there are many examples of philosophers who tried to use alleged biological facts—that is, "human nature"—to underscore their political arguments. But there are very few, if any, examples of working biologists who used their expertise to elaborate a large, complicated, and partisan vision of politics. In this respect Gould is very unusual, and so his system of thought is unusually interesting.

This book is not a biography of Stephen Jay Gould. It is an exploration in the history of ideas. I intend to analyze, explicate, and evaluate Gould's scientific and political ideas the way he tried to present them: as a coherent whole. Because many of his positions in both realms were unorthodox, I will have to elaborate some esoteric concepts that normally are not discussed in the same forum. The discussion may therefore become abstruse, but it is going to be fun.

By "politics" in the title of this book, I mean two things. First, I mean the internal politics of science, which, while they are fought out at an elevated level of discourse, present most of the characteristics of politics elsewhere, including personal attack, demagoguery, misrepresentation, and interests masquerading as principles. Second, I mean the politics of evolution in the wider society, which, in the United States, inevitably includes the attacks of modern creationists on the theory of natural selection. Until his death, Gould was deeply involved in both kinds of politics. This book is about his participation in the politics of both realms.

The M Word

I was auditing Professor Dan Bolnick's graduate course in "Speciation" at the University of Texas in early 2006. During the first class session, Professor Bolnick had introduced me as a Department of Government professor to the six graduate students in the course, and he had explained that I was writing a book on Gould. I happened to arrive ten minutes early to the second class session, and as I entered the room I found a single male student in his twenties, sitting at the central table and reading one of the professional papers that had been assigned for that day. As I walked in, he looked up and greeted me. I said good morning and sat down. In a friendly manner, he asked, "So, you're writing a book about Stephen Jay Gould?" When I acknowledged that I was, his eyes narrowed slightly, and he exclaimed, with a hint of amused irony in his voice, "Communist biology!"

That single expostulation encapsulates much of the hostility that surrounded Gould's writings within his own and related professions. Was Gould a Marxist? And were his theories disguised attempts to transform evolutionary biology into a propaganda vehicle for Marxist theories of contemporary society?

On the face of it, the theory of natural selection does not lend itself to Marxist interpretations. As originated most importantly by Charles Darwin, and secondarily by Alfred Russel Wallace in 1858, the theory rests on three primary postulates, which together force an inference. First, because every organism's environment is ruthlessly dangerous, all organisms produce more offspring than can possibly survive—hedging their genetic bets, as it were. Second, individual organisms vary in structure, function, and (in the case of animals) behavior. Third, the variabilities are inheritable, so that, as an inference, the environment will winnow those that conform less well to the demands of life, thus leaving the better-conforming (the "fit") to pass on their characteristics to the next generation.[14] As environments change over deep time, the constant culling and inheriting combine to make species slowly evolve into other species. All species, therefore, are descended from previous species, and the history of life is a gigantic genealogy. Beginning in 1900, evolutionary biologists began to marry this theory to the findings of Mendelian genetics, producing by the 1940s a "modern synthesis" that retained the original Darwin-Wallace outlines while specifying the methods by which heritable qualities were originated and passed on.[15]

In contrast, Marxism is a theory that, although its details have been interpreted in differing ways by different adherents, rests on the assumption that the history of human societies is at bottom the history of economic class conflict.[16] Since, of the millions of species of plants, animals, fungus, algae, and bacteria that have inhabited the Earth over more than three billion years, only one has ever generated economic classes, and since Darwinism (I follow the universal practice of referring to the theory as "Darwinism" rather than "Darwin-Wallaceism") deals with all of those species, it would seem to be a poor candidate for Marxist influence. Yet, as the expression of my graduate student acquaintance illustrates, the suspicion that Gould's work is somehow Marxism with a biological gloss has colored the reception of his work since the mid-1970s. It is a disciplinary folklore that needs to be evaluated.

In 1972, as ambitious young paleontologists, Gould and Niles Eldredge wrote a paper that rocked paleontology, and eventually had repercussions far beyond natural science. In "Punctuated Equilibria: An Alternative to Phyletic Gradualism" they argued that the traditional Darwinian model (also the model of the modern synthesis) of the transition from one species to another as a relatively slow, steady progression through infinitely small intermediate steps was not supported by the fossil record. Instead, they asserted, the history of life preserved in the rocks showed that species tended to appear rather suddenly (in geological time, which is still extremely slow by human time standards) and then persist with some oscillation of form, but no major morphological changes, until they went extinct. Rather than "phyletic gradualism" in the history of species, the fossil record showed long periods of "stasis" punctuated by geologically brief spurts of speciation.[17]

I will discuss in detail the theory contained in this article, and the reaction it engendered within evolutionary biology, in chapter 3. Here, the subject is Gould's supposed political allegiances and the way they allegedly contaminated his work. In 1977, he wrote an article for the journal *Paleobiology* in which he reviewed the evidence published since 1972 for and against the model of punctuated equilibrium, and attempted to clarify some of the conceptual issues it had left unresolved. In a concluding section, he also tried to explain how he and Eldredge had come to interpret the fossil record differently than most other people in their profession. In this section, Gould introduced one of his favorite themes—"that even the greatest scientific achievements are rooted in their cultural contexts."[18]

Darwin's phyletic gradualism, Gould argued, was situated in his personal position as an upper-class Englishman of the early nineteenth century. Preferring slow economic and political change in his society to revolutionary upheaval, he saw that sort of change in the history of life. Further, "the general preference that so many of us hold for gradualism is a metaphysical stance embedded in the modern history of Western cultures; it is not a high-order empirical observation, induced from the study of nature."[19] How, then, were Eldredge and Gould able to perceive the story in the rocks from a different metaphysical stance? Gould could not speak for Eldredge, but he suggested that his own mind was open to a new interpretation because he grew up in a distinct cultural milieu, a milieu that lived in opposition to the dominant intellectual trends of his society: "It may . . . not be irrelevant to our personal preferences that one of us learned his Marxism, literally at his daddy's knee."[20]

This statement is plainly an attempt to explain why one mind might be more receptive to the theory of punctuated equilibrium than another. It acknowledges the structural similarities between the Marxist theory of social stasis followed by revolution and the Eldredge-Gould theory of morphological stasis followed by speciation, and suggests that a mind familiar with the first might be prepared to recognize the second within the empirical record. It is in no sense an admission that punctuated equilibrium is Marxist biology. Furthermore, although Gould actually coined the famous phrase, the original concept underlying the first article was Eldredge's. Eldredge had not learned any Marxism at his daddy's knee; his personal background was Baptist and Republican.[21] Although he considered himself a political liberal, he was not close to being a Marxist. He had come to reinterpret the history of life not through any political discussion, but by working on the pattern of trilobite evolution during his PhD dissertation.[22] He had, in other words, as a good scientist been persuaded by the empirical evidence that a new model of evolution was needed. Marxism was irrelevant.

But no matter. The legend that "punctuated equilibria" was merely one conceptual brick in Stephen Jay Gould's edifice of Marxist biology became a professional truism that dogged Gould until the end of his career. No matter how he tried to explain himself, the legend spread, becoming one of those collective assumptions that need not be written down because it is part of the folklore of the discipline. In summarizing Gould's thought as Communist biology, Dan Bolnick's graduate student was repeating one of the myths of the tribe.

Since this book is an effort to explicate Gould's thought, however, I must address the issue head on. Was Gould a Marxist, and was his system of thought one version of Marxist biology? The answer to both questions is somewhat ambiguous, because Gould never actually sat down and wrote an essay of the "This I Believe" variety. He preferred to write on specific topics. I have put his topical writing together to see a large system, but that system is nowhere described in its entirety by Gould himself.

After Gould's death in 2002, biologist Richard Lewontin and ecologist Richard Levins, two avowed Marxists, wrote an appreciation of his life and work for *Monthly Review.* Not only were these two familiar with his writings, but Lewontin had cotaught a course at Harvard with Gould for many years, and he considered Gould a good friend.[23] "Despite our close comradeship in many things over many years, we never had a discussion of Marx's theory of history or political economy," they wrote. Their own assessment was that Gould was a "radical," a far more inclusive designation than "Marxist."[24]

Gould was on the advisory board of the journal *Rethinking Marxism,* the title of which suggests that he was no more an orthodox Marxist than he was an orthodox Darwinist. He was also associated with the Brecht Forum, sponsor of the New York Marxist School, and Science for the People, a loose group of left-wing scholars especially active in the attack on sociobiology described in chapter 4. Plainly, then, Gould was a leftist. But it is a dangerous form of ideology-by-association to ascribe a set of beliefs to a person because he or she belongs to a club. Anyone who has ever joined an academic organization knows that they are full of contentious individuals and riven by doctrinal quarrels—and this goes double for leftist groups. It is far safer and fairer to consult the actual published views of any given person under discussion.

I cannot claim to have read every single published work by Gould, nor every one of his hundreds of professional papers, nor the various book introductions, chapters, prefaces, and reviews he wrote for others. Nor have I had access to his private papers. I do claim to have read all his books, all his important (oft-cited) professional articles, and a good sample of his auxiliary publications. I have not found a single sentence in which he identified himself as a Marxist. On the contrary, there are several instances in which he distanced himself from Marxism, as when he wrote in 1995 of the "failure" of "Marx's theory of historical stages toward a communist ideal,"[25] or in 2002 that his political beliefs were "very different from my father's."[26] Furthermore, when he did occasionally summarize his politics, he tended to characterize himself using the capacious term "modern liberal,"[27] or the even more inclusive label "humanist."[28] Finally, in some of his asides he professed beliefs that would have coexisted uncomfortably alongside Marxist ideology, as when he wrote in 1991, "We live in a capitalist economy, and I have no particular objection to honorable self-interest,"[29] or when he publicly advocated a détente between science and religion.[30]

If Gould was not an orthodox or even unorthodox Marxist, what does it mean to say that his beliefs were on the left? The terms *left* and *right,* like the terms *liberal* and *conservative,* have tended to change over the years in the United States, and, even within that unstable environment, differ from the use of the terms in other democracies. To take one historical example, the Whig party, right wing in conventional usage, was in favor of government activism during the era prior to the Civil War, while the Jacksonian Democratic party, left wing in conventional usage, was virulently antigovernment. Today, the polarities are reversed, with the left-wing Democrats being more progovernment. Political scientists and historians have produced volumes of research on the definitions of left and right in popular understanding. I cannot add to this literature here, except to point out that it is filled with ambiguity and equivocation. There is no timelessly accurate description of ideology.

Nevertheless, if left and right do not adhere continuously to a set of beliefs, they do tend to associate themselves with a hierarchy of values. As summarized by philosopher Peter Singer, leftists are "on the side of the weak, not the powerful; of the oppressed, not the oppressor, of the ridden, not the rider," all of which, in practice, add up to an emphasis on *equality* as the defining value of the left.[31] There can be warm disagreements among leftists about what kind of, and how much, equality they should endorse. Gould was famous for arguing on many occasions some version of the position, "Human equality is a contingent fact of history."[32] He thus often proved himself a leftist. But what kind of leftist?

Among leftists, two great divisions can be found, corresponding to those who support government activity to ensure *equality of opportunity*—moderates in today's politics—and those who prefer government activity to enforce *equality of result*—the more extreme, including Marxists. Perhaps surprisingly, given his reputation, Gould's writings position his own system of beliefs within the moderate camp.

Although Gould often wrote, directly or indirectly, against such creators of social inequality as racism, sexism, and class discrimination, he did not support a complete equalization of rewards in society. He was quite clear in endorsing equality of opportunity but not result. "We too often, and tragically, confuse our legitimate dislike of elitism as imposed limitation with an argument for leveling all concentrated excellence to some least-common denominator of maximal accessibility," he wrote, and continued,

> Elitism is repulsive when based upon external and artificial limitations like race, gender, or social class. Repulsive and utterly false—for that spark of genius is randomly distributed across all the cruel barriers of our social prejudice. We therefore must grant access—and encouragement—to everyone. . . . We will have no justice until this kind of equality can be attained. But if only a small minority respond, and these are our best and brightest of all races, classes, and genders, shall we deny them the pinnacle of their soul's striving. . . . What is wrong with this truly democratic form of elitism?[33]

Within the welter of contradictory and ambiguous positions of contemporary political discourse, his endorsement of equality of opportunity but not result thus drew Gould away from the Marxist end of the scale, and positioned him firmly under the label he himself selected: modern liberal. As we will see, Gould's writings contained contradictions, puzzles, and extensions that both muddy and amplify his basic ideological position. Nevertheless, as a first characterization, it is most accurate to speak of Gould as a non-Marxist leftist.

In a sense, however, this conclusion is not interesting. I have had to clean out the suspicion of *sub rosa* Marxism in order to allow myself to address his actual beliefs, which are indeed fascinating. Before I can launch an explication of Gould's substance, however, I must address two other aspects of his person: his religious background and his rhetorical style.

The J Word

Both sets of Gould's grandparents were Eastern European Jews who immigrated to the United States in the first decade of the twentieth century. As a Marxist, Gould's father kept a secular household but retained a pride in his Jewish identity. Gould wrote about his childhood, "I shared the enormous benefit of a respect for learning that pervades Jewish culture."[34] Like his father, he maintained a secular outlook (describing himself as an agnostic)[35] but kept a strong identity as a Jew. He occasionally mentioned his cultural heritage in his essays.

By temperament I am inclined to think that a scientist's religious background is not germane to his or her theories and analyses. But one sometimes comes across serious scholars who believe otherwise. Michael Ruse, for example, has speculated that Gould's consistent opposition to the application of biological theories to human behavior is partly the result of his Jewish identity.[36] And there are two reasons to think that there is something in Ruse's suggestion. First, as mentioned in the previous section, Gould himself always insisted, in both his historical sketches of scientific personalities and in his professional papers, that all science, being produced in a cultural context, is best understood by acknowledging that context. Second, he publicly asserted that his opposition to the careless use of mental testing, and by extension, the application of evolutionary theory to human behavior, was at least partly motivated by the attempt of early intelligence testers to exclude Jewish immigrants by labeling them inferior to old-stock Americans, and by the murderous use that the Nazis made of eugenics against Jews.[37]

Nevertheless, once we have admitted that his religious background undoubtedly sensitized Gould to certain issues, I think that there is no explanatory power remaining in the fact that he was a Jew. Neither his scientific nor his political arguments were made as a Jew but rather as a scientist and as an American citizen. The scientists who argued with him never accused him of making points that were erroneous because they came from a Jew, and he never rebutted their criticisms by accusing them of arguing like gentiles. Everyone involved in the controversies appealed to the secular standards of reason and evidence; while the verbal fight more than occasionally degenerated into ad hominem attacks, it never extended to the ethnic groups of which the contestants were members. While Gould collaborated with Jews on some projects (Richard Lewontin, for example), he also collaborated with people of Christian background (Niles Eldredge, for example). Moreover, his appeal was universal, just as the appeal of E. O. Wilson, of Southern American Protestant background, and Richard Dawkins, of English Protestant background, were and are universal, and for the same reason: because all three were creative thinkers and terrific writers. I conclude (and I do not believe that the fact that I was raised by Protestant parents in Los Angeles has anything to do with it) that my original impulse to ignore Gould's Jewishness in writing about his ideas was the correct one, and I will do so for the rest of this book.

About his writing style, however, there is much more to be said, here and throughout this book.

Style Makes the Manuscript

Reading through many of the attacks and critiques that have been showered on Gould and his ideas over the decades, it is impossible not to receive the impression that some of the hostility was prompted by simple envy of his fabulous success. Would that we all could sell as many books and receive as many awards as this professor! Yet inside the envy there seems to be another, far more interesting and ideologically more important assumption providing the motivation for the envy itself: the belief that Gould made himself America's evolutionist laureate by rhetorical cheating.

Science is a method and an attitude, but both are accompanied by an expositional style. Or at least they are supposed to be. In the self-concept of natural scientists, they write with clarity and dispassion and never use tricks of rhetoric to persuade an audience when the facts are insufficient to accomplish that purpose. As Brooks and Warren articulated the "windowpane theory" of scientific discourse in 1938:

> The primary advantage of the scientific statement is that of absolute precision. . . . Such precision . . . can be gained only by using terms in special and previously defined senses. The scientist carefully cuts away from his scientific terms all associations, emotional colorings, and implications of attitude and judgment.[38]

Or as Davida Charney described the model of scientific prose in 1993:

> Since at least Bacon's time, scientists have taken as their ideal a form of scientific discourse that is straightforward, objective, and dispassionate, discourse that confines itself to describing independently confirmable observations and drawing dispassionately logical conclusions from them. As such, generations of scientists have conceived of their discourse as standing outside the realm of rhetoric, the classical art of persuasion.[39]

There are two problems with this ideal. The first is that in its pretension to be nonrhetorical, much scientific prose is disingenuous. While it may be logical, based on empirical analysis, and fair minded in its consideration of differing points of view, it is still fashioned as persuasive discourse. Its purpose is to lead the reader to a specified conclusion. Therefore, there is no nonrhetorical scientific prose, only better and worse rhetoric. As Gregson and Selzer demonstrate in their discussion of an essay by eminent evolutionary biologist John Maynard Smith (and incidentally, one of Gould's chief critics within the profession), Smith

> created a conventional, almost stereotypically scientific reader—an objective, fair-minded, sober, impersonal, careful, and reasonable pursuer of scientific truth. . . . Smith's reader in that circumstance was asked to assume a position of respectful inferiority to him: Maynard Smith was in effect an implied schoolmaster lecturing to willing students.[40]

Maynard Smith's rhetorical denial of his own rhetorical purpose was thus a pose. The pose did not make him a bad scientist or a poor writer—in truth, he was a greatly accomplished scientist and a clear writer. But the fact that Maynard Smith was himself a rhetorician does undercut the criticisms that he and others made of Gould as somehow treading outside the boundaries of appropriate discourse by writing in a more engaging style. Gould saw himself, correctly, as using the art of rhetoric to advance the cause of rationalism. In this book I will not always side with Gould's position on issues, but on this particular point I think he is dead right. Gould may have been mistaken in some of his notions about evolutionary theory, but on the issue of rhetoric he is in the clear.

The second problem with the "nonrhetorical" ideal of scientific prose is that it is manifestly contradicted by some of the most important scientific treatises of history. Darwin's *On the Origin of Species*, to take the founding document of evolutionary biology, is superbly rhetorical in its use of metaphor, its marshaling of vivid supporting detail, and its easy readability. And Gould, who rejected absolutely the Baconian ideal of the antiseptic research report, delighted in introducing his readers to fine scientific writing that contradicted the myth. He asserted, in general, scientific papers "are polite, self-serving fictions"[41] and that "science self-selects for poor writing."[42] But the exceptions are not just adequate prose; they are superb literature. Charles Lyell, whose 1830 to 1833 *The Principles of Geology* was extremely influential in persuading Darwin to adopt the uniformitarian (Eldredge and Gould's "phyletic gradualism") perspective on earth's history, was "a great writer, and much of his enormous success reflects his verbal skills."[43] Gould endorsed Sir Peter Medawar's judgment of D'Arcy Thompson's 1942 *On Growth and Form* as "beyond comparison the finest work of literature in all the annals of science that have been recorded in the English tongue."[44] Sigmund Freud, Gould averred, "rose to pre-eminence as a paramount social force through his unparalleled literary gifts, and surely not for his cockamamie and unsupported theory of the human psyche."[45]

Undoubtedly, here is one reason why Gould raised so many hackles among other biologists. He did not merely evade the Baconian ideal of scientific rhetoric; he publicly repudiated it. "Scientists," he wrote in 1993, "simply do not acknowledge that the form and language of an argument (as opposed to its logic and empirical content) could have anything to do with its effectiveness."[46] But as the examples of Lyell, Thompson, and Freud testify, this belief is a woeful misunderstanding. Therefore, Gould ignored the canons of the antiseptic myth and wrote as interestingly as he could. "I do what I do entirely on purpose and with definite aim," he said.[47]

> Above all, I believe in the commingling of the arts and sciences as the two greatest expressions of human creativity. I regret the petty and parochial boundaries that both domains have established—the impenetrable and sterile language of so much scholarship in the humanities, the dry, impersonal and barbarous passive voice of scientific prose.[48]

And so, exactly what did Gould do on purpose and with definite aim? How can we all become charming writers? These turn out to be difficult questions to answer. Good writing is a little like Justice Potter Stewart's celebrated attitude toward pornography—we cannot define it, but we all know it when we see it. Even the scholars of rhetoric, for all their ability to analyze and categorize communication, cannot explain why some writing is charming and some is dull or offensive.

Take, for example, a prominent textbook, *Modern Rhetorical Criticism*, by Roderick Hart. This 374-page compendium explores the concepts and methodologies used by scholars of rhetoric to study their subject, from "the five basic moves" of rhetoric to the "logic of persuasion," to "prominent" vs. "passive authorial status" in communication, and much more.[49] It contains a variety of useful ways to categorize, think about, and perhaps resist, attempts to persuade. Yet when it comes to the core of the problem—the differentiation of persuasion that is certain to move an audience from persuasion that is ignored or rejected—the author must admit that "despite centuries of interest in rhetorical style, it remains elusive. . . . Even for non-Catholics, the special friend of the stylistic critic is St. Jude, the patron saint of the hopeless."[50]

Nevertheless, even without enumerating clear rules of charming composition, rhetorical research provides some tantalizing hints into Gould's success. For example, Hart reports on a 1985 study by David Sinclair on "successful" Southern Baptist preachers (those whose congregations were growing) versus the "unsuccessful" variety (whose congregations were contracting). The growing congregations, Sinclair found, "heard sermons that were more personal, more narrational, more assured, and more detailed than parishioners in the declining churches."[51] Without doubt, Gould's scientific essays were more personal and narrational than those of his colleagues, and possibly more detailed. When it comes to persuasion, there may be less difference between the religious and the scientific communication styles than one would expect.

Two of Gould's coauthored articles have been specifically analyzed by scholars for rhetorical content—the aforementioned essay on punctuated equilibrium, which was treated in an article in the *Quarterly Journal of Speech*, and a 1979 polemic against adaptationism (to be discussed in chapter 4) written with Richard Lewontin, which rated an entire book-length treatment by sixteen scholars.[52] Although the authors spoke of the "inter-audience movements" in the article, and brought to bear the conceptual resources of dramatism, structural linquistics, politeness theory, stasis theory, the "thinking aloud" method, the "reader-response" perspective, social constructionism, intertextual and citation analysis, the cultural studies approach, deconstruction, and feminism, none of them were able to explain why Gould is so much fun to read.

Having received only limited support from the professionals, I am going to try my amateur hand at analyzing Gould's charm. I can identify no single key to understanding his compositional method. Instead, I will list the elements that seem to enter into a typical Gould essay. He identified some of these elements himself. Others are the patterns I have noticed in his writing.

Scientific Witnessing

Gould likes to bring the reader along on a voyage of discovery. He does not so much present us with a conclusion we are to accept, as admit us to his own experience and permit us to participate vicariously in the process of realization. "Here is what I learned about X, and how I learned it, and the broader implications of my new knowledge," might be a paradigmatic summary of many of his essays. By letting the reader in on the hunt for knowledge, Gould creates what rhetorical scholars call "identification" between himself and his readers. The members of the audience join emotionally with the author and exult in his conquest of truth.[53]

Personal Reminiscence

Gould's prose is, therefore, not only scientific, but unabashedly personal. As part of the strategy of relating personally to the audience, he frequently uses his own experiences, scientific or otherwise, to introduce or otherwise facilitate his theme. Some sentences from the first paragraphs of a variety of his essays will give the flavor of this personalizing technique:

> "When I was 10 years old, James Arness terrified me as a giant, predaceous carrot in *The Thing*."[54]
>
> "I once had a gutsy English teacher who used a drugstore paperback called *Word Power Made Easy* instead of the insipid fare officially available."[55]
>
> "As my son grows, I have monitored the changing fashions in kiddie culture for words expressing deep admiration—what I called 'cool' in my day, and my father designated 'swell.'"[56]
>
> "I grew up in New York and, beyond a ferry ride or two to Hoboken . . . never left the city before age ten. But I read about distant places of beauty of quiet, and longed to visit the American West."[57]

Even at the beginning of his career Gould was a more personal writer than most scholars. But he became intensely personal over time, and late in life expressed dislike for his first book of essays, *Ever Since Darwin*, published in 1977, because "I now find these essays too generic in lacking the more personal style that I hope I developed later."[58]

Situating Evolutionary Biology in a Cultural Context

It seemed to be Gould's assumption that the more interconnections between evolutionary theory and the wider cultural world he could establish, the more his arguments would seem comfortable, familiar, and therefore less threatening to his readers. Narrow specialization is off-putting, but cultural contextualization seems to include the reader in a conspiracy of wisdom. So Gould's essays were a cornucopia of relevant erudition. He worked in the views of philosophers (Kuhn, Popper, Foucault, Karl Marx, Peirce, Hume, Hobbes, Russell, and, as a favorite villain, Spencer), economists (Smith, Malthus), lyricists (W. S. Gilbert, Julia Ward Howe), poets (Pope, Tennyson, Omar Khayyam, Virgil, Wordsworth, Coleridge, Milton, Dryden, Blake), painters (Michelangelo, Rubens), composers (Rossini), writers (Stevenson, Poe, Shakespeare, Steinbeck, Nabokov, Kipling), psychologists (Freud, Jung), and comedians (Groucho Marx, Charlie Chaplin), as well as often alluding to passages in both testaments of the Bible. He also frequently brought in two subjects that interested him, architecture (especially European cathedrals) and baseball. None of these quotations and allusions were inserted gratuitously; all helped to further the points that Gould was attempting to make.

Gould's critics, of course, had a different view of his habit of cultural contextualization. One of the standard features of a Gould essay, harrumphed John Alcock, was "the ostentatious display of erudition."

> I suspect that by salting his articles with these items he persuades many a reader that he is an erudite chap, one whose pronouncements have considerable credibility. . . . By advertising his scholarly credentials, Gould gains a debater's advantage, which comes into play when he contrasts his erudition with the supposed absence of same in his opponents.[59]

Alcock was correct, but in a sense the criticism was not pertinent. His real quarrel with Gould was over the substance of evolutionary theory. The fact that Gould employed what Alcock considered to be rhetorical tricks would not have been of concern if Alcock had agreed with Gould on substance. Besides, there is something churlish in the complaint that your opponent is a better writer than you are.

Informality

Like cultural contextualization, informality of style breaks down the barriers between readers and what is, in fact, a rather esoteric subject matter. The world of laboratories, statistics, and methodological quarrel is obviously alien to the nonscholar. Not quite so obviously, differences of outlook and method create intellectual—and thus psychological—barriers even between members of the same discipline. By writing as informally as possible, even in most of his professional publications (*Ontogeny and Phylogeny* is the great exception), Gould managed to familiarize an unfamiliar subject and make it seem less forbidding.

Gould often repeated some version of his claim (this one from *Won-*

derful Life) that "the concepts of science, in all their richness and ambiguity, can be presented without any compromise, without any simplification counting as distortion, in language accessible to all intelligent people."[60] Whether he succeeded in conveying truth without distortion is a matter of contention within his discipline, but there is no doubt that he tried to make his points in homespun, even vulgar words and phrases that would bring the subject down out of the clouds:

"Factor analysis, rooted in abstract statistical theory and based on the attempt to discover 'underlying' structure to large matrices of data, is, to put it bluntly, a bitch."[61]

"Charles Darwin developed a radical theory of evolution in 1838 and published it twenty-one years later only because A. R. Wallace was about to scoop him."[62]

"Perhaps the Grim Reaper of anatomical designs is only Lady Luck in disguise."[63]

"Nature, apparently, can make a gorgeous hexagon, but she cannot (or did not deign to) make a year with a nice even number of days or lunations. What a bummer."[64]

Similarly, Gould often combined his informality of style with an informality of subject matter. He did not necessarily need to draw on organisms with Latin names for a fund of examples. He seemed able to find pertinent lessons about evolutionary theory in many of the familiar entities of popular culture. Thus, one of his early essays analyzed the "evolution" of the image of Mickey Mouse: "The abstract features of human childhood elicit powerful emotional responses in us, even when they occur in other animals," he wrote. It appears that the cartoon image of the world's most famous mouse underwent a process of "neotony"—Mickey looked younger as time went on. And so, "Mickey Mouse's evolutionary road down the course of his own growth in reverse reflects the unconscious discovery of this biological principle by Disney and his artists."[65]

In like manner, Gould wrote serious analyses of the evolution of the Hershey Bar[66] and the decline of the .400 hitter in baseball.[67] Although British biologist Richard Dawkins complained that the references to America's national pastime in the latter were incomprehensible,[68] they obviously resonated in the larger country's culture.

Metaphor

Scholars of rhetoric have paid much attention to metaphor—an idea, generally visual, that symbolically compares one thing to another. Some maintain that ordinary human language would be impossible without metaphorical expressions.[69] Others argue that "phenomena become objects of scientific discourse by virtue of the metaphor that makes them accessible to cognition."[70]

Not all scientists have agreed with the last position. Newton referred to metaphor as "a kind of ingenious nonsense," and Bacon described it "rather as a pleasure or play of wit than a science."[71] But there is no doubt that metaphors play a prominent role in evolutionary science. Darwin's fame as a prose stylist partly rests on his facility with metaphor: the tangled bank, the tree of life, the wedge, the struggle for life, selection. And modern evolutionary biologists almost seem to compete in their eagerness to invent new metaphors that may become part of the imagination of the discipline. Professional publications contain such tropes as the "the blitzkrieg theory,"[72] the "Swiss-army knife model,"[73] "the Lazarus effect,"[74] the "house of cards" model,[75] the "snowball effect,"[76] the "Red Queen" model,[77] "arms races,"[78] "selfish genes,"[79] and many more.

Gould was perhaps the most enthusiastic creator of metaphors in the history of the science. "I love metaphors," he told us, "I use them all the time." He did not just love them; he had well-articulated reasons for using them:

> We need carriers, or metaphors, to make . . . imaginative jumps. Moreover, a scholar's choice of metaphor usually provides our best insight into the preferred modes of thought and surrounding social circumstances that so influence all of human reasoning, even the scientific modes often viewed as fully objective in our mythology.[80]

Gould not only liked to use his own metaphors—as a historian of science he was fascinated by the ways others used them. He devoted several of his essays to exploring the implications of other scientists' choice of metaphors. The subject of his book *Time's Arrow, Time's Cycle* was the way metaphorical depictions of time have led to differing views of the history of life.

But anyone so sensitive to the use of metaphors must also be aware of their power to mislead as well as enlighten. Indeed, Gould spent almost as much effort warning against the metaphors he disliked as in elaborating his own. "Metaphor is a dangerous, if ineluctable, device," he told us. "We use images and analogies to foster understanding . . . but we run the risk of falsely infusing nature with the baggage of our parochial prejudices or idiosyncratic social arrangements."[81] As we shall see in the coming chapters, much of the ideological warfare in which Gould engaged consisted of attacking the favored metaphors of his adversaries and defending his own images.

Dichotomy

Gould often argued against this or that trend in thinking among evolutionists. When doing so, he tended to simplify and exaggerate the differences between those who agreed with him (the good guys) and those who didn't (those who were badly mistaken). In some of his essays, the world of biology was divided into two camps, the larger composed of scientists whose ontology and epistemology blinded them to the nuances of truth, and the smaller, standing with Gould, who had wised up.

Naturally, the people portrayed as being in the myopic camp did not like the label. As Alcock objected, one of the "basic elements" of a "Gouldian polemic was the erection of a strawman to attack," and another was "the effort to claim the moral high ground."[82] Nobody likes to be criticized in public, especially if one's views are simplified and perhaps distorted. Indeed, Gould himself was known to complain that other writers misrepresented his opinions and to assert that "nothing depresses me more than the rampant, seemingly inveterate mischaracterization that lies at the core of nearly every academic debate."[83]

But whether Alcock was right that Gould used dichotomization to distort or Gould was right that his own views were being distorted, or both, neither had a unique right to complain. Unfair dichotomies, it appears, are a feature of all scientific controversy. As philosopher and historian of science David Hull summarized his own researches:

> The tendency to read more homogeneity into a group than actually exists, is all but impossible to avoid. . . . Conversely, scientists who find themselves in rival camps go to the opposite extreme and exaggerate how much they disagree, an exaggeration that is further enhanced by differences in terminology.[84]

But Hull also counseled us not to worry. Whether Gould or anyone else misrepresented other scientists' views by dichotomizing, it would all work out for the best. "(D)istortion is more than routine in science," he insisted. "It is a traditional mode of argumentation, and a mode that is not entirely counterproductive. It forces scientists to commit themselves."[85]

Empathy

In contrast to his sometimes rough treatment of living people with whom he disagreed, Gould often showed remarkable respect and sympathy for the dead. In his historical writings he frequently went over familiar ground, yet he managed to fashion startling new insights by combining close reading of familiar sources, sensitivity to historical context, and a willingness to reevaluate the seemingly settled.

A prime example was his resurrection of William Jennings Bryan. Along with Bishop Wilberforce in England, Bryan, the thrice-defeated presidential candidate, who in 1925 attempted to prosecute Tennessee schoolteacher John Scopes for daring to introduce Darwinist thinking to his students, stands as one of evolution's hissable villains. The image of Bryan as the symbol of intellectual reaction was cemented by Fredrich March's portrayal of the fading politician as a self-righteously intolerant hick in the 1955 film *Inherit the Wind.* In the mythology of evolutionism, Bryan embodies the forces of darkness that are always trying to suppress scientific thought.

Gould, however, asked us to cast aside our devotion to myth and do some thinking about historical context. Darwinism, he reminded us, had for decades been appropriated by some of the most repellent political forces in the United States and the world. "Social Darwinism," as espoused by Herbert Spencer in England and William Graham Sumner in the United States, had been preaching that people on top of society were, by that very fact, superior, and that the people on the bottom were similarly getting what they deserved. Moreover, German militarists had appropriated Darwin as a justification for their predatory intentions. It was plausible, Gould argued, that a progressive on all other issues such as Bryan would have seen evolutionary theory as an ally of reaction. He quoted one of Bryan's letters: "I learned that it was Darwinism that was at the basis of that damnable doctrine that might makes right that had spread over Germany."[86] Although a proper interpretation of the theory would have led Bryan and everyone else to understand that

"Darwinism implies nothing about moral conduct,"[87] given the fact that many bad people were inferring the opposite, Gould suggested that "Bryan had the wrong solution, but he had correctly identified a problem!"[88] Suddenly Bryan was de-demonized and portrayed as a person deserving sympathy and understanding. In like manner, many of Gould's other essays asked us to take another look at historical figures we thought we understood

From the Particular to the General

Small details, from the peculiar shape of a flower petal to the horrifying life-cycles of some parasites, have fascinated the science oriented for many centuries. But the details, without a unifying conceptual framework, have no intrinsic meaning. Likewise, theory alone, without illustrative examples, may be of interest to specialists but is rightfully boring to the uninitiated. It was Gould's genius to be able to simultaneously illustrate the big principles with vivid details and make the details meaningful with reference to the principles.

"You have to sneak up on generalities, not assault them head-on," he wrote. "I have come to understand the power of treating generalities in particulars."[89] And so to make the essential point that "the proof of evolution lies in imperfections that reveal history,"[90] he spent most of his essay not on the principle itself, but on a story about the panda's thumb.[91] To refute the creationist charge that the fossil record lacks intermediate forms, he reported on the recent discoveries of four transitional species between land mammals and whales.[92] To illustrate his theme that evolution does not necessarily involve change toward greater complexity, he discussed physical simplification among *Sacculina*, a group of parasites that infest crabs and their relatives.[93]

It is not true that God (or the devil) is in the details. It is true that understanding lies in the interplay between large principles that render the world comprehensible and tiny bits of reality that embody the principles. Gould managed to occupy an intermediate position between "popularizers" who write about disconnected facts and scholars who can only convey arid theory. The result was charming, and educational.

The Serious Popularizer

Gould wanted his readers to understand that although he admitted to having literary pretensions, he was not to be thought of as only a scientific popularizer. He asked to be seen as a working scientist who wrote accessibly about science. This self-concept was so important to Gould that he told his readers he often put "original findings" into his popular essays, and he requested that scholars cite them as readily as they referenced his articles in scientific journals. "In scholars' jargon, I hope and trust that my colleagues will regard these essays as *primary rather than secondary sources*."[94] In this book I will honor his request and treat all his writings, regardless of the forum in which they appeared, as equivalently indicative of his thought.

Tensions

Charming and educational though he might be, Gould was not necessarily correct in all of his opinions. Although he was a paleontologist by profession, that career mixed him in among the much larger association of scientists who go by the general description "evolutionary biologists." He was thus a scientist who studied evolution, and as such, he agreed with virtually all other scientists about most of the contours of life's history. He vigorously opposed the apostles of irrationality, whether they called themselves "scientific creationists" or advocates of "intelligent design," in their efforts to cast doubt on the basic concepts of Darwinism and in their efforts to insert the teaching of superstition into American public schools. In that sense, Gould stood with all other scientists.

On the subtleties of evolutionary theory, however, Gould was an outlier. He was not a lone eccentric, pushing his own personal view of life's history. But he was the most famous representative (although not in every case the leader, and not necessarily, on every issue, the most powerful thinker) of a minority group within natural science who believed that Darwinist theory as envisioned by most evolutionary biologists was mistaken. Furthermore, he was a member of a minority group within the minority group who saw political implications to all the scientific disagreements within the larger profession. Thus, Gould had a series of arguments with majority opinion in evolutionary biology that proceeded along two tracks simultaneously: the purely scientific and the political. In very brief summary, here are the questions about which Gould and his allies quarreled with the majority of other people in evolutionary biology:

1. Does evolution of species proceed by insensible change over vast periods of time, or are species relatively stable for long periods, then, faced with environmental crisis, do they suddenly (within the parameters of deep time) give rise to other species?
2. Similarly, are species simply the form of any given genetic line at any particular time (that is, are they snapshots in a continuum), or are they discrete entities with a definable existence? Are we to view species as gene frequencies or as discrete organic entities?
3. Is the pattern of the transformation of many species over time (macroevolution) to be explained with reference to the extrapolation of tiny changes (microevolution) over geologic time, or is macroevolution a fundamentally different process that must be studied with separate concepts?
4. Is the history of evolution a set of basic processes repeated endlessly, or has it been subject to random, large-scale interruptions?
5. To what extent should scientists emphasize adaptation to the environment as an explanation for the form and function of each organism, and to what extent should they emphasize the constraints that limit possible adaptations?
6. What, exactly, are the entities that are evolving? Are they genes, or organisms, or species, or higher groupings of organisms?
7. Is there a direction to evolution? That is, are living things becoming more complicated, or more diverse, or more intelligent, or more anything? And if there is a direction, what is it and what causes it?
8. If the answer to the above question is yes, can we predict life's direction?
9. Once we have learned about the history of life in general, what can we say about the history of the species *Homo sapiens*? Is anything we learn from the history of that species relevant to understanding the individual and social problems we experience today?
10. Do the answers to any of the above questions have political implications for society at large? If so, what are the implications, and how

do they arise out of the scientific argument? And if they do have political implications, what can we do about it?

In the rest of this book I will explore these questions, Gould's answers to them, and the arguments he got into with others who disagreed. But to anticipate the answer to question number ten: a word that recurs in this book is "implication." Sometimes, an argument over the politics of biology is a straightforward clash of opinions about some methodology or research conclusion. Much more often, however, the disagreement is about the implications of some means employed by some scientists, or the equally controversial implications of the ends they are pursuing. Gould spent his last three decades fighting about the implications of his theories about life's history, the implications of his opinions about the proper way to study human behavior, the implications of his preferred definition of human intelligence, the implications of his opposition to the adaptationist paradigm, the implications of his insistence on the importance of contingency in evolution, and so on. Historians and philosophers of science have tended to ignore the political implications of means and ends, preferring to focus on the strictly scientific issues confronting their subjects. I think that to focus on only the scientific conflicts without discussing their political implications, however, would be a misleading way to study the thought of Stephen Jay Gould. A large portion of this book, therefore, is a set of discussions about the political implications of scientific ideas.

notes

1. Stephen Jay Gould, *The Panda's Thumb: More Reflections on Natural History* (New York: W. W. Norton, 1980), p. 267.

2. Stephen Jay Gould, *Hen's Teeth and Horse's Toes: Further Reflections on Natural History* (New York: W. W. Norton, 1983), pp. 281, 313; Stephen Jay Gould, *Time's Arrow, Time's Cycle: Myth and Metaphor in the Discovery of Geological Time* (Cambridge, MA: Harvard University Press, 1987), p. vii.

3. Stephen Jay Gould, *Dinosaur in a Haystack* (New York: Harmony Books, 1995), p, 229.

4. Stephen Jay Gould, *An Urchin in the Storm: Essays about Books and Ideas* (New York: W. W. Norton, 1987), p. 78.

5. Ibid., p. 195.

6. Gould, *Time's Arrow*, p. 18.

7. Stephen Jay Gould, *The Structure of Evolutionary Theory* (Cambridge, MA: Harvard University Press, 2002), p. 1105.

8. From the flyleaf of ibid.

9. Cornelia Dean, "How Quantum Physics Can Teach Biologists about Evolution," *New York Times*, July 5, 2005, p. D2.

10. Daniel Dennett, *Darwin's Dangerous Idea: Evolution and the Meanings of Life* (New York: Simon and Schuster, 1995), p. 266.

11. Ibid., pp. 262–312.

12. John Alcock, "Unpunctuated Equilibrium in the *Natural History* Essays of Stephen Jay Gould," *Evolution and Human Behavior* 19 (1998): 321–36.

13. J. Philippe Rushton, "Race, Intelligence, and the Brain: The Errors and Omissions of the 'Revised' Edition of S. J. Gould's *The Mismeasure of Man*," *Personality and Individual Differences* 23, no. 1 (1997): 169–80.

14. Charles Darwin, *On The Origin of Species By Means of Natural Selection* (New York: Barnes and Noble Classics, 2004 [1859]).

15. Ernst Mayr and William B. Provine, eds., *The Evolutionary Synthesis: Perspectives on the Unification of Biology* (Cambridge, MA: Harvard University Press, 1998).

16. Karl Marx and Friedrich Engels, "The Communist Manifesto," 1848, in Arthur P. Mendel, ed., *Essential Works of Marxism* (New York: Bantam, 1965): pp. 13–44.

17. Niles Eldredge and Stephen Jay Gould, "Punctuated Equilibria: An Alternative to Phyletic Gradualism," in Thomas J. M Schopf, ed., *Models In Paleobiology* (San Francisco: Freeman, Cooper and Company, 1972), pp. 82–115.

18. Stephen Jay Gould and Niles Eldredge, "Punctuated Equilibria: The Tempo and Mode of Evolution Reconsidered," *Paleobiology* 3 (1977): 145.

19. Ibid.

20. Ibid., p. 146.

21. Niles Eldredge, in discussion with the author, June 26, 2006.

22. Niles Eldredge, *Time Frames: The Evolution of Punctuated Equilibria* (Princeton, NJ: Princeton University Press, 1985), pp. 55–56, 70–71.

23. Richard Lewontin, in discussion with the author, June 21, 2006.

24. Richard C. Lewontin and Richard Levins, "Stephen Jay Gould: What Does It Mean to Be a Radical?" *Monthly Review* 54, no. 6, November, 6, 2002. http://www.monthlyreview.org/1102lewontin.htm, accessed November 27, 2005.

25. Gould, *Dinosaur in a Haystack*, p. 346.

26. Gould, *Structure of Evolutionary Theory*, p. 1018.

27. Gould, *Urchin in the Storm*, p. 179; Stephen Jay Gould *I Have Landed: The*

End of a Beginning in Natural History (New York: Three Rivers Press, 2003), p. 219.

28. Stephen Jay Gould, *Leonardo's Mountain of Clams and the Diet of Worms* (New York: Three Rivers Press, 1998), pp. 2, 4.

29. Stephen Jay Gould, *Bully for Brontosaurus: Reflections in Natural History* (New York: W. W. Norton, 1991), p. 101.

30. Stephen Jay Gould, *Rocks of Ages: Science and Religion in the Fullness of Life* (New York: Vintage, 1999).

31. Peter Singer, *A Darwinian Left: Politics, Evolution, and Cooperation* (New Haven, CT: Yale University Press, 2000), pp. 8–9.

32. Stephen Jay Gould, *Flamingo's Smile: Reflections in Natural History* (New York: W. W. Norton, 1985), p. 186.

33. Gould, *Dinosaur in a Haystack*, p. 246.

34. Gould, *Rocks of Ages*, p. 8.

35. Ibid.

36. Michael Ruse, *Mystery of Mysteries: Is Evolution a Social Construction?* (Cambridge, MA: Harvard University Press, 1999), p. 145.

37. Stephen Jay Gould, *The Mismeasure of Man*, 2nd ed. (New York: W. W. Norton, 1996), dedication, pp. 38–39, 197, 255, 258; Gould, *I Have Landed*, p. 173.

38. Cleanth Brooks and Robert Penn Warren. *Understanding Poetry* (New York: Holt, Rinehart, and Winston, 1938), p. 4, quoted in Jack Selzer, ed., *Understanding Scientific Prose* (Madison: University of Wisconsin Press, 1993), p. 4.

39. Davida Charney, "A Study in Rhetorical Reading: How Evolutionists Read 'The Spandrels of San Marco,'" in Selzer, *Understanding Scientific Prose*, p. 203.

40. Gay Gregson and Jack Selzer, "The Reader in the Text of 'The Spandrels of San Marco,'" in Selzer, *Understanding Scientific Prose*, p. 185.

41. Gould, *Urchin in the Storm*, p. 85.

42. Gould, *Time's Arrow*, p. 107.

43. Ibid.

44. Gould, *Panda's Thumb*, p. 40.

45. Stephen Jay Gould, *The Hedgehog, the Fox, and the Magister's Pox* (New York: Three Rivers Press, 2003), p. 134.

46. Stephen Jay Gould, "Fulfilling the Spandrels of World and Mind," in Selzer, *Scientific Prose*, p. 323.

47. Ibid., p. 321.

48. Ibid., p. 322.

49. Roderick P. Hart, *Modern Rhetorical Criticism*, 2nd ed. (Boston: Allyn and Bacon, 1997), pp. 7, 84, 214.

50. Ibid., pp. 133, 154.

51. Ibid., p. 165.

52. John Lyne and Henry F. Howe, "'Punctuated Equilibria'; Rhetorical Dynamics of a Scientific Controversy," *Quarterly Journal of Speech* 72 (1986): 132–47.

53. Hart, *Rhetorical Analysis*, pp. 27, 275.

54. Gould, *Since Darwin*, p. 113.

55. Gould, *Hen's Teeth*, p. 107.

56. Gould, *Bully for Brontosaurus*, p. 309.

57. Stephen Jay Gould, *Eight Little Piggies: Reflections in Natural History* (New York: W. W. Norton, 1993), p. 220.

58. Gould, *I Have Landed*, p. 4.

59. Alcock, "Unpunctuated Equilibrium," pp. 322–23.

60. Stephen Jay Gould, *Wonderful Life: The Burgess Shale and the Nature of History* (New York: W. W. Norton, 1989), p. 16.

61. Gould, *Mismeasure of Man*, p. 238.

62. Gould, *Since Darwin*, p. 21.

63. Gould, *Wonderful Life*, p. 48.

64. Stephen Jay Gould, *Questioning the Millennium* (New York: Harmony Books, 1999), p. 163.

65. Gould, *Panda's Thumb*, pp. 97, 104.

66. Gould, *Hen's Teeth*, pp. 313–19.

67. Gould, *Flamingo's Smile*, pp. 215–29.

68. Richard Dawkins, *A Devil's Chaplain: Reflections on Hope, Lies, Science, and Love* (New York: Houghton Mifflin, 2003), pp. 206–207.

69. Hart, *Rhetorical Analysis*, p. 146.

70. Richard Harvey Brown, "Rhetoric and the Science of History: The Debate between Evolutionism and Empiricism as a Conflict of Metaphors," *Quarterly Journal of Speech* 72 (1986): p. 149.

71. Newton and Bacon quoted in ibid., p. 148.

72. David M. Raup, *Extinction: Bad Genes or Bad Luck?* (New York: W. W. Norton, 1991), pp. 89–93.

73. Peter Carruthers and Andrew Chamberlain, "Introduction," in Carruthers and Chamberlain, eds., *Evolution and the Human Mind: Modularity, Language, and Meta-Cognition* (Cambridge: Cambridge University Press, 2000), pp. 1–12.

74. John Alroy, "Constant Extinction, Constrained Diversification, and Uncoordinated Stasis in North American Mammals," *Palaeo* 127 (1996): p. 293.

75. D. Waxman, and S. Gavrilets, "20 Questions on Adaptive Dynamics," *Journal of Evolutionary Biology* 18 (2005): p. 1148.

76. H. Allen Orr, "The Population Genetics of Speciation: The Evolution of Hybrid Incompatibilities," *Genetics* 139 (April 1995): 1805–13, esp. 1812.

77. Richard Dawkins, *The Blind Watchmaker: Why the Evidence of Evolution Reveals a Universe without Design* (New York: W. W. Norton, 1996), p. 183.

78. Richard Dawkins, *The Selfish Gene* (Oxford: Oxford University Press, 1976), p. 250.

79. Dawkins, *Selfish Gene*.

80. Gould, *Dinosaur in a Haystack*, p. 444.

81. Ibid., p. 433.

82. Alcock "Unpuctuated Equilibrium," pp. 324, 325.

83. Gould, *Eight Little Piggies*, pp. 45, 125.

84. David L. Hull, *Science as a Process: An Evolutionary Account of the Social and Conceptual Development in Science* (Chicago: University of Chicago Press, 1988), p. 13.

85. Ibid., p. 289.

86. Bryan quoted in Gould, *Bully for Brontosaurus*, p. 422.

87. Ibid., p. 427.

88. Ibid., p. 429.

89. Gould, *Mismeasure of Man*, p. 20.

90. Gould, *Panda's Thumb*, p. 13.

91. Ibid., pp. 21–24.

92. Gould, *Dinosaur in a Haystack*, pp. 361–65.

93. Gould, *Diet of Worms*, pp. 356–71.

94. Gould, *I Have Landed*, pp. 6–7.

Chapter 2

philosophy of science

What is true? It seems upon first glance that science is the best approach humans have invented to answer that question with tolerable accuracy. During the last century and a half or so, science, and its allied disciplines medicine and engineering, have transformed the face of the globe and the conditions of human life. The great prestige that science receives from the public at large has been earned with results. The general, if seldom-articulated attitude among scientists and nonscientists alike often seems to be that if it works so well, it must be true.

Nevertheless, closer inspection reveals that truth in science is more ambiguous than it first appears. As scientists themselves have often written, they do not think much about the philosophical foundations of their work; they are too busy participating in the enterprise to reflect upon its ultimate justification. Physicist Steven Weinberg's observation, "We learn about the philosophy of science by doing science, not the other way around,"[1] has been echoed by many other scientists, including Gould. And simply doing science does not require much philosophy, as long as things are going well.

As it happens, however, things often do not go well in science. When scientists are forced to explain to themselves and others why what they say is to be believed, they tread on the territory of professional philosophy. They enter, first, the domain of ontology. In a general sense, this is the branch of metaphysics concerned with identifying the nature of reality at its most basic. In a more focused and practical sense, when scientists choose ontological commitments, they are adopting, implicitly or explicitly, positions about entities, forces, and the ways they interact. When anyone asserts that the universe is created by human thought, for example, he or she is staking an ontological position. Second, scientists are forced to endorse one among several possible theories of epistemology, the field of philosophy that investigates the nature and possibility of human knowledge. Scientific positions on methodology always rest on epistemological assumptions.

Scientists must become philosophers under either of two sets of circumstances. The first is when pseudosciences such as astrology, intelligent design, and New Age theories attempt to borrow some of the prestige of science for their own uses. When the imposters are at the door, scientists are forced to ask themselves, "How can we discern an imposter from a real scientist?" Among philosophers this puzzle is called the "problem of demarcation."[2] It is in the nature of philosophers to equivocate, and during the twentieth century the discipline split into several camps that offered clashing solutions to the problem of demarcation, or in some cases denied that it could be solved. Scientists, observing no consensus among philosophers, have tended to settle on a simple and basic definition of genuine as opposed to fraudulent science that emphasizes the confrontation of theory with empirical evidence, as in this summary by biologist John A. Moore:

> Science is a way of knowing by accumulating data from observations and experiments, seeking relationships of the data with other natural phenomena, and excluding supernatural explanation and personal wishes.[3]

In other words, virtually all scientists, as opposed to many philosophers and most pseudoscientists, are realists and empiricists. That is, they believe that the universe exists independently of human thought or observation and that the way to understand reality is to test each others' theories against evidence.[4] Gould would agree with this formulation. However, in the actual activity of science the ideal of empiricism often must be modified or reinterpreted to take account of numerous ambiguities. As a group, scientists rarely bother to confront these ambiguities. Gould was one of the few who took the time not just to do science but to participate directly in the philosophy of science and to thus work out a position on numerous issues within its boundaries.

The second and more common situation in which philosophy becomes relevant to the activity of science is when different individuals or groups within science formulate incompatible theories to explain the same natural phenomena. Even within a generally dominant research program—and Darwinism is a powerful example here—there are schools, rebels, methodological preferences, differing interpretations of evidence, clashing world-views, and incompatible political preferences. When scientists argue with each other, they are often forced to appeal to the standards of their craft to justify their favored point of view. These arguments inevitably include controversies over what the standards are, or should be. Sometimes the appeals address fundamental issues. Often the appeals are ad hoc and verbal; occasionally they are long considered and published.

As one of the most enthusiastic participants in the controversies that racked evolutionary biology from 1972 to 2002, Gould paid much more than his share of attention to the philosophy of science. He did not come to this interest late; he had been a dual geology/philosophy major in his undergraduate years.[5] Sometimes his essays and professional articles directly addressed some disputed issue. More often, a section of a paper on a substantive point included attention to some relevant philosophical controversy. His philosophical writings are noteworthy for three characteristics. First, like almost all of his work, they tended to be very well written. Second, he had read and assimilated many of the controversies within the philosophy of science, but he had not been able to resolve them all. His thought thus contained contradictions and hedges. Third, he was always aware of the moral and political implications of sometimes even very esoteric positions. For Gould, philosophy was never just about science. It was also about the larger political context in which science operates. He was attempting to understand the world, but his ultimate goal was always to change it for the better.

Provisional Truth vs. Relativism

The twentieth-century overthrow of Newtonian physics inaugurated by Einstein in 1905 shook the intellectual world within and beyond science. In a society that often seemed to be threatened by moral ambiguity, classical physics had appeared be the one indisputably reliable truth. To scientists, and nonscientists who followed intellectual progress, "the" scientific method stemming from Newton was the place to take an epistemological stand. Morality, to the nonreligious, might be relative and irreducibly ironic, but physical reality was secure.

With Einstein, and with the rise of quantum mechanics during the 1920s, some of the confidence of science dissolved. The title of a biography of Nobel Prize winner Max Born nicely captures the change in world-view wrought by early twentieth-century physics: *The End of the Certain World*.[6] Although physics remained enormously prestigious, it, and science in general, suffered a considerable cognitive demotion among intellectuals. Ontological doubt entered the very concept of science. If Newton's laws, reliably unequivocal yesterday, were exposed as relative today, what might happen to the new generalizations tomorrow? If sincere, thinking people had believed classical physics to be true when in fact it was just a limited case of larger principles, how could anyone now put faith in those new principles? On what basis could anyone say that science was true?[7]

The twentieth century produced two traditions of philosophy to try to answer these questions. Although they both were sometimes inconsistent and unclear, occasionally changed their minds, and evolved in their ideas over time, the century's two most influential philosophers of science, Karl Popper and Thomas Kuhn, were rough embodiments of the two traditions. I will treat each of them as exemplars by simplifying their thought. I justify this simplification not only because discussing individuals is easier than discussing whole traditions but because Gould also treated them as exemplars. Although he was not deeply read in the whole tangled corpus of twentieth-century philosophy of science, he frequently cited Popper and Kuhn when discussing his own ideas. He never publicly acknowledged the fact that the philosophy of each pointed in a different direction.

Kuhn's solution to the problem of the relativity of scientific knowledge was to embrace it.[8] In each era, as he argued in his 1962 classic, *The Structure of Scientific Revolutions*, scientists subscribe to a "paradigm," which is not clearly defined but seems to be a mixture of world-view, theory of knowledge, and research program. When working within "normal science," practitioners share a paradigm (Newtonian physics, for example). But eventually, empirical puzzles arise that the reigning paradigm cannot handle (difficulties with the observed behavior of light). At some point there is a crisis, and a new paradigm arises (the theory of relativity, then quantum mechanics) that handles the old problems as well, and the new problems better, than the old paradigm. In a "scientific revolution" the old is discarded and the new adopted.[9]

If Kuhn had stopped there, his book would have been useful but not disturbing. In a late chapter, however, he put forward a thesis that, although Kuhn hedged it with many qualifications, had clearly radical implications. Having adopted a new paradigm, he wrote, scientists "practice their trades in different worlds. . . . We may . . . have to relinquish the notion, explicit or implicit, that changes in paradigms carry scientists and those who learn from them closer and closer to the truth"[10]

Although Kuhn denied, in the first edition of his book and for years afterward, that he meant to say that *reality* changes when paradigms change, he seems not to have realized the power of his implied argument. Despite his qualifications, it is difficult to read his book without concluding that he believes there is no reality but only shared images of reality. Intended or not, that idea, unendorsed by its author, took off and became a major component of Western philosophy. Soon other philosophers, sometimes identified as "externalists,"[11] were elaborating the notion that science actually bears only a tangential relation to reality and instead reflects individual and shared psychological needs and ontological assumptions. In particular, a large tradition in modern philosophy assumes that all scientific theory is "socially embedded," or, in other words, a reflection of the (largely economic) needs of the time and place of the scientist rather than an objective portrait of the universe.[12]

Gould was strongly influenced by Kuhn (the title of his own magnum opus, *The Structure of Evolutionary Theory*, may have been an unconscious homage), and by the extension of his theory into ontological relativism. "If I were to cite any one factor as probably most important among the numerous influences that predisposed my own mind" toward cocreating the theory of punctuated equilibrium, he wrote, "I would mention my reading, as a first-year graduate student in 1963," of Kuhn's book.[13] He cited Kuhn numerous times, not only in his discussions of the punctuational theory, but in support of his oft-repeated assertion that all theories must be examined with sensitivity to their social context. After mentioning Kuhn's book as a milestone in *Time's Arrow, Time's Cycle*, he added,

> Science may differ from other intellectual activity in its focus upon the construction and operation of natural objects. But scientists are not robotic inducing machines. . . . Scientists are human beings, immersed in culture, and struggling with all the curious tools of inference that mind permits. . . . Objective minds do not exist outside culture, so we must make the best of our ineluctable embedding.[14]

Is Gould, then, an "externalist," who believes that all science is a reflection of social or individual forces rather than a window to reality? No, he is in fact an "internalist" who believes in objective truth.[15] In this he follows Popper, whose reaction to the Newtonian meltdown was not to abandon realism but to attempt to fashion a set of rules that would allow science to provisionally approach reality.

"Although there are no general criteria by which we can recognize truth—except perhaps tautological truth—" Popper explained, "there are something like criteria of progress towards the truth."[16] Popper's central criterion, which he expanded upon for decades, was the notion of *falsifiability*. All theories worthy of scientific respectability must generate predictions that could, at least in principle, fail empirical tests. Theories must be repeatedly tested against empirical facts. Theories that fail are to be considered revealed as false, although new, modified versions can be created to be tested again. As long as they pass the tests, theories are not to be considered true, but merely provisionally acceptable. Nevertheless, a theory that has passed many tests may be considered to have been "corroborated" by them, and may be used with some confidence by scientists.[17]

Gould cited Popper often, and with approval. Reality exists, and may be approached with provisional faith. "Notwithstanding a long history of arguments, ranging from the playful to the tendentious to the sophistic, there is a world out there full of stars, amoebas, and quartz crystals," he told his readers in *Eight Little Piggies*.

> Science does construct better and better maps of this outer reality, so we must assume that change in the history of scientific theories often records more adequate knowledge of the external world, and may therefore be called progress.[18]

As with Popper, Gould emphasized repeated empirical testing of theories.[19] He endorsed the idea that the "potential fallibility" of scientific claims is the "only universal attribute of scientific statements. . . . If a claim cannot be disproven, it does not belong to the enterprise of science."[20] And he spent a considerable amount of time amassing and evaluating empirical evidence relevant to his and Eldredge's theory of punctuated equilibria.[21]

There were, then, two sides to Gould's brain, the Kuhn side and the Popper side, and the two were in tension. He believed that all knowledge was socially grounded, and therefore that reality was contingent, dependent, and problematical. But he also believed that reality was objective and that science could approach it to ever-closer proximity. Not being stupid, he recognized the tension and attempted to reconcile the two sides.

The key to the attempted reconciliation was Gould's contention that the "social embeddedness" of observation and theory could be liberating rather than constraining. "Science is, and must be, culturally embedded;" he conceded, "what else could the product of human passion be?" But,

> Science is also progressive because it discovers and masters more and more (yet ever so little *in toto*) of a complex external reality. Culture is not the enemy of objectivity but a matrix that can either aid or retard advancing knowledge. Science is not a linear march to truth but a tortuous road with blind alleys and a rubbernecking delay every mile or two.[22]

To illustrate the liberating, creative side of cultural embeddedness, Gould offered his favorite example in all things: Darwin. The master did not pluck his theory of natural selection from the assembled facts of nature he witnessed on the *Beagle*. He combined them with knowledge he had gained from his reading in fields outside natural history. In particular, Darwin took the theory of the wholesome effects of market competition that he found in Adam Smith and Thomas Malthus, transmuted the setting of the concepts, and applied them to the history of life. Thus, natural selection was "the economy of Adam Smith transferred to nature."[23] If Darwin had not had the classical economists to inspire him, Gould implies, he might never have come up with his grand theory. Darwin's embeddedness in the culture that produced and circulated classical economic ideas permitted his own mind to generate a new theory.

Gould was therefore what philosopher Marjorie Grene termed a "comprehensive realist," meaning that he believed that the world existed outside his perceptions, but was nevertheless conscious of the fact that his perceptions and preconceptions colored his observations and cognitions.[24] I cannot say, however, that I find his attempted reconciliation of Kuhn and Popper convincing. In the end, Gould's version of comprehensive realism does not offer us a way to choose between internalist and externalist approaches to science. Is Darwin's theory true despite its grounding in classical economic theory? How do we know, if we ourselves are culturally embedded? Gould glided around these questions, never quite addressing them squarely. Instead, he fashioned a compromise between the alternatives that allowed him both to do science and to write about it, but that did not directly confront the hard ontological issues. From a practicing scientist, this may have been the only expedient strategy. From a philosopher, it left the impression of clever evasion.

Experimental vs. Narrative Science

Until roughly the 1970s, thinking about the philosophy of science was distorted by everyone's preoccupation with physics. As the most successful of the sciences, and the one whose prestige had been most directly challenged by Einstein, physics was held up not only as a model to which all other sciences should aspire, but as the science that generated the most interesting philosophical problems. When Popper and Kuhn discussed science, for example, almost all their examples were drawn from physics, with a few from chemistry thrown in.

The emphasis on physics had two important consequences for other sciences, both baleful. In the first place, practitioners of the other sciences—I include both biology and the social sciences here—struggled for decades to live up to an inappropriate ideal. As Mohan and Linsky summarized physics' appeal, it "lends itself to elegant axiomatization . . . its laws are universal and exceptionless, and seemingly they are statable without reference to dubious concepts such as causality."[25] Philosophers of science explained to nonphysicists that if they wanted to be real scientists they should strive to create "covering laws" from which predictions about the future could be deduced, and test those predictions in the laboratory.[26] Because biology, not to mention social science, contained few laws that were universal and exceptionless, and was meaningless without the concept of causality, it floundered for decades in a vacuum of justification.[27]

During the 1970s, however, both philosophers of science (Elliott Sober, Michael Ruse, Alex Rosenberg, Daniel Dennett) and biologists themselves (Ernst Mayr, John A. Moore, Gould) began to elaborate a philosophy of science that was specifically appropriate to evolutionary biology.[28] As with all science, there were many disagreements of general approach and specific application among the various theorists, although Gould tended to agree on many points with Mayr, one of the authors of the modern synthesis and his colleague at Harvard.

The second baleful consequence of the physics fixation was a feeling of inferiority among evolutionary biologists (a feeling once again shared by social scientists). Although, as a timeless generalization about the workings of nature, natural selection is a concept comparable to gravity, it is not directly measurable. The effects of gravity were being measured informally by humans long before Newton was born. The effects of natural selection, however, must be inferred, and the process is anything but obvious. Further, because the environment is different for every organism in every situation, the outcome of the process of selection cannot be known ahead of time; it is not predictable in the same sense that the orbit of a comet can be anticipated far into the future. As a result, while physics and chemistry got the glory and the big government grants, evolutionary biology was relegated to an inferior status. This situation infuriated Gould, who often tried to convince his audience not only of its injustice, but of its inaccuracy:

> Several years ago, Harvard University, in an uncharacteristic act of educational innovation, broke conceptual ground by organizing the sciences according to procedural style rather than conventional disciplines within the core curriculum. We did not make the usual twofold division into physical versus biological, but recognized . . . the experimental-predictive and the historical. We designated each category by a letter rather than a name. Guess which division became Science A, and which Science B? My course in the history of earth and life is called Science B-16.
>
> Perhaps the saddest aspect of this linear ranking lies in the acceptance of inferiority by bottom dwellers, and their persistent attempt to ape inappropriate methods that may work higher up the ladder. When the order itself should be vigorously challenged, and plurality with equality asserted in pride, too many historical scientists act like the prison trusty who, ever mindful of his tenuous advantages, outdoes the warden himself in zeal of preserving the status quo of power and subordination.[29]

In making his argument that science was "the greatest of human adventures,"[30] Gould never failed to emphasize that the historical sciences (cosmology, geology, evolutionary biology) were every bit as worthy of partaking in disciplinary prestige as were the more experimental sciences such as chemistry and physics. Although "I am a historian at heart," and although "paleontologists are . . . historians by heart and by profession,"[31] the historical method was not inferior to the experimental method in achieving provisional knowledge. The methods of the two differ in a superficial manner.

Physicists and chemists are able to work with their empirical subject in real time, manipulate it, and observe changes in effects as they alter the causes. The historical sciences, on the other hand, rely upon the use of "consilience of induction"[32] applied to evidence created in the past to fashion a "narrative explanation"[33] that allows scientists to infer "the processes we cannot see from results that have been preserved."[34]

This historical method of doing science, while different in kind from the experimental method, did not yield results that were in any respect inferior. "Historical science is not worse, more restricted, or less capable of achieving firm conclusions because experiment, prediction, and subsumption under invariant laws of nature do not represent its usual working methods. The sciences of history use a different mode of explanation, rooted in the comparative and observational richness of our data,"[35] but a mode that is completely equal in its ability to yield understanding. In particular, historical sciences rely upon the "primary methodological criterion of testability"[36] to just as great an extent, and are just as open to the possibility of falsification.[37] They search just as hard for ways to quantify their evidence.[38] Therefore, they share with experimental sciences the fundamental and scientifically defining claim to be able to provisionally access the "causal consequences of spatiotemporally invariant laws."[39]

Testing in the historical sciences is somewhat different than testing in a laboratory. Technically, the method used should be called "postdiction"[40] rather than prediction. Normally, in physics and chemistry, scientists predict what would be observed if they arranged reality in a particular way (shot electrons into a block of uranium, for example), then perform the operation and see if the prediction comes true. Historical science, in contrast, involves postdicting something that has already happened (or not happened) based upon inference from present data, then checking to see if the expectation of the past is met. Gould recounted his own version of the method, putting the Popperian criterion of falsifiability into a typically informal statement:

> Every time I collect fossils in Paleozoic rocks (550 to 225 million years old), I predict that I will not find fossil mammals—for mammals evolved in the subsequent Triassic period. . . . If I find fossil mammals, particularly such late-evolving creatures as cows, cats, elephants, and humans, in Paleozoic strata, our evolutionary goose is cooked.[41]

Successful postdiction permits "narrative explanation"[42] "that explicitly links the understanding of outcomes not only to spatiotemporally invariant laws of nature, but also, if not primarily, to the specific contingencies of antecedent states, which, if constituted differently, could not have generated the observed result."[43] This contingent nature of natural history has two important consequences. The first I will discuss in the next chapter. The second is that historical scientists are usually limited to postdiction; the future cannot be scientifically anticipated because the circumstances, in terms of either the genetic configuration of the organism or the environmental situation, are never the same twice. As a consequence, Gould agreed with one of his mentors, George Gaylord Simpson, who wrote in 1964, "The search for historical laws is . . . mistaken in principle. Laws apply . . . in common parlance to 'other things being equal.' But in history . . . other things are never equal."[44] According to Gould, therefore, historical scientists should not expect to be able to produce precise postdictions, measurable to the millimeter, gram, or year. Instead, evolutionary biology mainly assesses "relative frequencies among numerous possible outcomes," evaluating the statistical occurrence of various patterns in the fossil record, rather than relying on clean and decisive crucial experiments.[45] In interpreting the past, paleontologists examine evidence that is inherently incomplete and frequently contaminated.

Granting Gould's point that historical sciences are not inferior to those modes of exploration that rely much more heavily on experiment, still, reliance on "relative frequency" does introduce indeterminacies into the study of the past that do not occur (or occur much less often) in the experimental sciences. The basic source of information for paleontology—the fossil record—is still extremely unsatisfactory a hundred and fifty years after *On The Origin of Species*. Paleontologists estimate that only about 10 percent of the organisms in a marine environment stand a chance of becoming fossilized, and a considerably smaller proportion of those that lived on land. Moreover, there are severe biases among those that are preserved and those that are not. It is the hard parts of creatures that are potentially preservable—teeth, bones, shells. Soft organisms are generally lost to observation.[46] For example, their common occurrence today suggests that earthworms have been around for a very long time. Yet because they have no hard parts, they are completely absent from the fossil record.[47] Even among species that possess "hard" parts, some are less hard, and therefore less preservable, than others. Because birds must fly and therefore are under selective pressure to save weight, for example, they have evolved very light, frail bones that do not fossilize well. As a result, the fossil record of the development of birds is skimpy.[48]

The systematically insufficient nature of their professional data has forced paleontologists to engage in controversy about some of their most basic understandings of life's history. For example, for most of the history of paleontology, scientists believed that there was a "Cambrian explosion" about 570 million years ago, in which life, having been exclusively unicellular for roughly 3 billion years, suddenly crossed the threshold into multicellular forms, resulting in an immediate proliferation into many structural types. But there are paleontologists who doubt this story. They argue that there was multicellularity aplenty before the alleged explosion; but because it consisted of creatures with only soft bodies, it has left almost no record. The alleged explosion of forms, they argue, is thus simply an artifact of the faulty evidence.[49] This and other controversies about the proper way to interpret the events of the Cambrian explosion, if there was an explosion, have been part of the conversation of evolutionary biology for decades. Gould has participated vigorously in these debates. In fact, as we shall see, his own preferred interpretation of the beginning of the Cambrian has infuriated many others in the profession, and sparked spirited rejoinders.

In science, the more indeterminate the data, the more important becomes the interpretive style of the individual scientist. When rival theories are being evaluated by the discipline, fuzzy patterns generated by an imperfect fossil record must be fit into this or that conceptual framework. Physicists and chemists have their disagreements, but, even taking into account the problems physicists are having finding evidence to apply to debates over string theory, their latitude for interpretive variance is smaller because their data is usually less ambiguous. As a consequence, paleontology and other historical disciplines give greater play to the politics of science. Organizational ability and rhetorical style become more important when the methods of evaluating data are less precise. For most of Gould's career, he defended the provisionally firm knowledge base of his discipline to the outside world, while engaging in a series of skirmishes with fellow evolutionary biologists that arose from the problematic and arguable nature of the evidence they shared.

Reductionism vs. Holism

When a biologist looks at any organism, say, a squirrel, she must choose a perspective with which to understand it. Shall she interpret the squirrel as a system of chemical reactions, or as a collection of genes, or as an independent entity within an environment, or as a dependent entity partially defined by its interactions with the environment, or as a cell within the local community of squirrels, or as a cell within the local ecology, or, perhaps, as part of a world that is itself alive? The way she defines the squirrel will have important implications for the methods she chooses to study it, for the way she interprets her data, and for the way she presents her findings to the other members of her discipline. Some scientists would argue that there is no "correct" way to define the squirrel; it depends on the nature of the research question. Others would contend that there is, in fact, only one right way to define a squirrel, because the wrong definition involves misinterpretations that lead the scientist astray no matter what her research question.

In addition, there are a few within the scientific community who maintain that the way the squirrel is defined has more than *scientific* implications. Each way of looking at the world, they argue, also involves *political* implications. These scientists do not separate the scientific from the political; in their view, the right way of doing science is correct in both domains, and the wrong way is both scientifically misleading and politically reprehensible.

Gould's friend, colleague, and coauthor, molecular biologist Richard Lewontin, has been among the most vocal of scientists in arguing this view that the scientific and political cannot be separated. Gould and Lewontin do not agree on every issue, and on some issues it is not clear whether they agree. I cannot, therefore, simply quote Lewontin's views as if they were Gould's. Nevertheless, on many points of philosophy, Gould's published opinions are simply more cryptic versions of Lewontin's extended treatments of the relevant issues. On those topics I will use Lewontin's publications (many of which are coauthored) to help me explain some of Gould's own views.

In 1985, Lewontin published a book entitled *The Dialectical Biologist*, coauthored with population biologist Richard Levins. The purpose of this book is to argue that "a bourgeoise ideology of society has been writ large in a bourgeoise view of nature" and that "the Darwinian theory of evolution is a quintessential product of the bourgeoise intellectual revolution."[50] The authors do not claim, in this book, that modern scientists consciously or unconsciously act as propagandists for the capitalists who dominate their societies. The argument is much more subtle. Instead, Levins and Lewontin urge us to understand that the ontological commitments that have undergirded science since Bacon and Descartes force scientists to think in a manner that is convenient for the continued wealth and power of elites.

The major ontological villain, in their scenario, is reductionism. Since Bacon and Descartes, Western scientists have assumed that the best way to understand anything, alive or not, is to break it up into the smallest pieces possible, analyze how those pieces function, then aggregate the functioning of the pieces to understand the functioning of the whole. As Levins and Lewontin put it, "It is an explicit *objective* of Cartesian reductionism to find a very small set of independent causal pathways or 'factors' that can be used to reconstruct a large domain of phenomena. . . . The error of reductionism as a general point of view is that it supposes the higher-dimensional object is somehow 'composed' of its lower-dimensional projections, which have ontological primacy and which exist in isolation, the 'natural' parts of which the whole is composed."[51]

A confusion is introduced here, for there is a second meaning to reductionism: it is the attempt to understand the behavior of the tiny parts, and by extension the behavior of the whole, according to the most basic and general scientific theory. The ideal of all reductionists is to be able to explain everything, including evolution, by the laws of physics and chemistry. But even if they never succeed in explaining evolution with references to the theories of these "hard" disciplines, biological reductionists aspire to be able to explain the large contours of evolution using principles addressed to the most basic and general particles of life, genes.

Few writers seem to have noticed that there are thus two definitions of reductionism: analysis into the smallest possible parts, and subsumption under the most general scientific laws. Scientists debating the issues of the appropriate use of the concept constantly conflate the two, and as a result it is not always easy to follow their arguments. In a practical sense the conflation does not matter, because virtually all scientists who support one kind of reductionism also support the other, and it is similar for those who reject reductionism. Nevertheless, if in the following discussion I do not always make clear what sort of reductionism I am discussing, it is not my fault—the confusion is in the primary material.

Levins and Lewontin argue that by perpetuating the radical individualism of capitalist society, reductionist science not only makes scientists systematically misunderstand the organisms and ecologies with which they work but also reinforces the false political ideology that sees human beings as individual actors rather than parts of a social whole. Cartesian science, therefore, is conservative science. "Dialectical," or left-wing science, in contrast, would both break out of the reductionist mindset and permit the conceptualization of humans as socially situated. Bad science is therefore bad politics, and a change to good science might help the cause of good politics.

Lewontin's view, therefore, expressed with Levins and his other coauthors, is that all scientists, everywhere, whether consciously aware of it or not, follow what Ullica Segerstrale has termed a "coupled agenda."[52] The correct ontological stance in science inexorably implies morally defensible political conclusions; incorrect ontology just as inescapably must lead to morally reprehensible politics. As Lewontin, Steven Rose, and Leon J. Kamin put it in 1984, "an assertion of the separation of fact from value, of practice from theory, 'science' from 'society' is itself part of the fragmentation of knowledge that reductionist thinking sustains and which has been part of the mythology of the last century of 'scientific advance.'"[53] And as Lewontin and Levins summarized it a year later, "Scientists, whether they realize it or not, always choose sides."[54]

Levins and Lewontin's recommendation, which is thus both scientific and political, is methodological holism:

> The first principle of a dialectical view . . . is that a whole is a relation of heterogeneous parts that have no prior independent existence *as parts*. The second principle, which flows from the first, is that, in general, the properties of parts have no prior alienated existence but are acquired by being parts of a particular whole.[55]

Holism's additional implication is that it prevents the behavior of organisms from being determined mechanistically. An assumption of reductionism is that, sooner or later, a level of understanding will be reached that will allow the scientist to predict everything: crank the right assumptions into the computer and the behavior of all the organisms in an ecosystem can be anticipated. This principle also applies, of course, to *Homo sapiens*. The scientist who strives to find a reductionist science of evolution, therefore, is hoping for a world in which organisms, including humans, are the prisoners of iron laws of nature. As Weinberg, a committed reductionist, puts it, "Determinism is logically distinct from reductionism, but the two doctrines tend to go together because the reductionist goal of explanation is tied in with the determinist idea of prediction: we test our explanations by their power to make successful predictions."[56]

On the other hand, a holistic scientist must grant some independence of behavior to all organisms, and especially to humans, because whole organisms are not simply the sum of their parts. They behave at a level above that which can be understood by the aggregation of physics and chemistry. In some sense, they make choices. So reductionism is allied with "biological determinism,"[57] and the two reinforce social forces that benefit from the "ideology of inequality" by "relocating the cause of inequality from the structure of society to the nature of individuals,"[58] while holism teaches us that "it is our biology that makes us free."[59]

Lewontin and his coauthors elaborate these principles into other areas of scientific practice. I will discuss some of the details and impact of this holistic philosophy of science in future chapters. Here, I want to point out that, in an empirical sense, it is simply wrong to assert that there is a necessary ontological coupling between a reductionist stance toward science and a left-wing, or "dialectical," political orientation. On the one hand, other leftists do not agree with Lewontin's critique of reductionism. Nobel Prize–winning French biologist Jacques Monod was a Marxist, yet he strongly endorsed reductionism in his discipline. The holists, he wrote in *Chance and Necessity*, make a "foolish and wrongheaded" argument, "merely testifying to the holists' profound misapprehension of scientific method and of the crucial role analysis plays in it."[60] Biologists John Maynard Smith and Richard Dawkins in England and Robert Trivers in the United States are well-known advocates of a reductionist, gene-centered point of view, yet all three have been active in left-wing politics in their personal lives.[61] On the other hand, Ernst Mayr was a strong critic of reductionism in evolutionary biology for decades, and the "beanbag genetics" associated with it. In many essays he elaborated on the point, writing that "attempts to 'reduce' biological systems to the level of simple physico-chemical processes have failed because during the reduction the systems lost their specifically biological properties."[62] Yet Mayr's voluminous writings show no evidence of political ideology of any kind. I conclude that there is no *necessary* connection between reductionism and right-wing politics, nor between holism and left-wing politics.

Nevertheless, Gould follows Lewontin in arguing that there is such a connection, although his position is more ambiguous and nuanced. "Leftist scientists are more likely to combat biological determinism just as rightists tend to favor this quintessential justification of the status quo as intractable biology: the correlations are not accidental," he writes, sounding very much like Lewontin.

> But let us not be so disrespectful of thought that we dismiss the logic of arguments as nothing but an inevitable reflection of biases. . . . We have campaigned against this doctrine because we regard determinist arguments primarily as bad biology—and only then as devices used to support dubious politics.[63]

In another forum he expands on the theme:

> One encounters flawed arguments by the score every day. A decision to grant special scrutiny and attention to such an argument may then be influenced by other factors—including dislike of its implications. . . . But the flaw is quite independent of the political implications.[64]

In other words, although bad science does produce bad politics, and vice-versa, there is no invariable ontological connection. Gould never actually rejects Lewontin's argument directly, but his own position does not include it. Acute readers are therefore left with the question: what causes the observed correlation between reductionist ontology and conservative politics? It is not clear if Gould simply wants to duck this question, or whether he believes that whereas there may be an isomorphism between politics and scientific ontology, it cannot be assumed but must be established anew with each issue, or whether, as a liberal rather than a Marxist he is uneasy with the ontological connection. In any case, although Gould's "coupled agenda" is similar to Lewontin's, he declines to connect it to the ontological choices of individual scientists.

Gould's crusade against reductionism became most vivid in his series of exchanges with the most famous reductionist in biology, Richard Dawkins. In 1976 appeared *The Selfish Gene*, Dawkins's lucid, rigorous, and enormously influential argument in favor of interpreting evolution as being a process in terms of which organisms serve their genes, rather than the other way around. Humans, he asserted, like "all other animals, are machines created by our genes."[65] The way to try to explain any natural history puzzle is to ask, how can we account for this structure or behavior with reference to the differential survival and replication of genes?

For example, Dawkins asks, why is it that some parasites are detrimental, even deadly, to their hosts, while others are relatively or wholly benign? He then exemplifies his proposed method of inquiry by suggesting that

> the most important question to ask about any parasite is this: Are its genes transmitted to future generations via the same vehicles as the host's genes? If they are not, I would expect it to damage the host, in one way or another. But if they are, the parasite will do all it can to help the host, not only to survive but to reproduce.[66]

Dawkins then offers some examples (supplemented in a later book)[67] that support the theory. His intellectual approach is thus reductionist in both senses of the term—he is trying to analyze evolution at the level of the smallest feasible entity, and at the level of the most general theoretical construct. The fact that the approach seems to work in an empirical sense, not only in the parasite example but in regard to many other conjectures about a variety of other types of other organisms, lends it a great deal of credibility.

Nevertheless, Gould adds his voice to those of Lewontin, Mayr, and the philosophers Elliott Sober, Philip Kitcher, David Hull, and others, who are not convinced.[68] Although this controversy is too convoluted to cover in detail, the most interesting and powerful argument that Gould makes is that because of the phenomenon of "emergence," or "nonadditive interaction," it is impossible to fix the locus of evolution at any one level of life. Under some circumstances, a gene-centered focus is appropriate. Under other circumstances, discovering the truth of the history of life may require a focus on organisms, or groups within a species, or whole species, or levels above the species.

The concept of emergence is best illustrated with the familiar example of table salt. Individually, sodium is a poisonous metal and chlorine a poisonous gas. But combine them and the result is sodium chloride, a compound exhibiting entirely different characteristics from each of its components, including usefulness as a condiment. This quality of the whole having a different existence from its component parts ensures that

> a higher unit may form historically by aggregation of lower units. But so long as the higher unit develops emergent properties by nonadditive interaction among parts (lower units), the higher unit becomes, by definition, an independent agent in its own right, and not the passive 'slave' of controlling constituents.[69]

That independent agent, then, given the right selectionist pressure, may become a unit of selection. The result will be the survival of some genes in some organisms, but to name the genes as the agents that caused the survival of the organisms is to reverse the historical causality. Gould thus maintains that Dawkins and other gene reductionists have "confused bookkeeping with causality"—that is, by focusing on the gene to the exclusion of other possible units of selection, they have mistaken the results of the evolutionary process for its engine.[70]

Dawkins training was in ethology, the study of animals' behavior in their natural habitats. It is not surprising, then, that his focus is on the present consequences of past evolution and that virtually all his examples are drawn from the study of living species. Gould, in contrast, has been trained in paleontology, the study of patterns of life and death over immense time spans. It is not surprising, then, that many of his examples are of extinct species. While Dawkins thus finds much support for his point of view among present-day organisms, Gould finds much support for his view among organisms long dead. I do not make this point in order to argue that scholarly training must predispose scientists to different ontological positions. My point is that, for the rest of us, it may be difficult to choose between alternate troves of empirical examples because the different types of evidence lend themselves to different types of inference.

Moreover, the choice of reductionist or nonreductionist explanations by each theorist is a function of the fact that Dawkins and Gould have differing opinions about the motivation for the enterprise. For Dawkins, the fundamental point of evolutionary biology is to explain adaptive complexity—to answer the questions of why there are so many varieties of starfish, so many types of mating displays among birds, so many strategies of reproduction, and so on. As he tells the reader in the preface to *The Blind Watchmaker*, "The problem is that of complex design. . . . The complexity of living organisms is matched by the elegant efficiency of their apparent design. . . . This amount of complex design cries out for an explanation."[71]

In contrast, for Gould the point of evolutionary biology is to explain and interpret the patterns of life's history. Have organisms become more complex over time, and as a result, is life now in some aggregate sense fitter than it was a billion years ago? Have the breaks and branchings observable in the fossil record resulted from some sort of evolutionary creativity within organisms, or are they the consequence of catastrophic disruptions from outside, in the environment? Is evolution very flexible, so that life somehow finds a way to adapt to the conditions in which organisms find themselves, or is it highly constrained, permitting only limited paths of development? When Gould tells his readers, "We are the offspring of history," and "Evolution is a narrative science," he is expressing his view that getting the grand, overall narrative vision right should be the goal of evolutionary scientists.[72]

Therefore, when Gould would offer empirical evidence in his essays, his examples were often chosen to address questions of historical pattern rather than questions of adaptive complexity. One of his most interesting discussions centered on the peculiar case of the surviving diatoms. It seems that the mass extinction at the end of the Cretaceous era about 65 million years ago—the one famous in pop culture as the death knell of the dinosaurs, thus giving mammals their chance—presented paleontology with many other cases of differential survival. For example, the giant dust cloud raised by the asteroid that slammed into earth was lethal to species in several genera of photosynthetic plankton. Seventy-three percent of the total species of the coccolithophorids disappeared; 85 percent of the radiolarian, and 92 percent of the foraminifera. Yet, "the diatoms, perhaps the most prominent group of photosynthetic plankton, sailed through the Cretaceous debacle with a generic loss of only 23 percent."[73] Why did the diatoms survive while their relatives died?

Gould did not bother to point out that Dawkins's genes'-eye-view would not allow us to address this issue. He left the lesson implicit: if we are to understand what happened during the years after the impact of the asteroid, we must analyze the situation in terms of types of organisms, not merely individuals, and certainly not clusters of genes. It seems that the diatoms as a genus shared certain characteristics that turned out to be useful:

> Diatoms evolved the capacity to form resting spores in order to wait out predictably fluctuating seasons of inhospitable environments. They developed this key adaptation for ordinary life in normal times, not in anticipa-

> tion of relative success should a comet strike the earth several million years in the future. Yet if months of darkness triggered death by extraterrestrial impact, then diatoms held a fortuitous leg up for survival.[74]

In its own way, this example is as convincing as Dawkins's discussion of parasites. The diatoms' survival must be understood as the result of adaptations shared among species at least two levels above that of the gene. In a technical sense, this story of differential survival could be told as a genes'-eye-view horror story: genes built diatom bodies for ordinary times, and those genes perished in extraordinary times. But at the level of the understanding of evolutionary patterns, the gene-centered approach is unsatisfying for this example. If we want to comprehend the history of life, we must ask why *types* of organisms died (formanifera, dinosaurs) while other *types* carried on (diatoms, mammals). To say that an asteroid hit the earth, and some distributions of genes survived while others did not, is to explain nothing that anyone wants to know.

The debate between reductionists and antireductionists has been a feature of biology since Darwin and will likely remain to trouble scientists when Gould and Dawkins are forgotten.[75] But, as we have seen, the debate has a political side also. Although Gould and Lewontin were persistent and resourceful in relating their ontological commitments to their political commitments, members of the audience witnessing the debate did not always pay attention to that side of their arguments. While some scientists reacted with something approaching hysteria to the political side of Gould and Lewontin, others, as well as some philosophers, acted as if they were embarrassed by the introduction of politics into biology, and seemed to express a desire to sweep the issue under the rug.

Thus, philosopher Kim Sterelny managed to write a useful book with a clever title, *Dawkins vs. Gould: Survival of the Fittest*, without giving more than brief and superficial attention to the political dimension of Gould's thinking.[76] Similarly, philosopher Timothy Shanahan's thoughtful and even-handed examination of controversies within evolutionary biology, *The Evolution of Darwinism*, gives many pages of analysis and evaluation to arguments between Gould and Dawkins, and Gould and other scientists, yet somehow manages to completely miss the relevance of politics to Gould's science.[77]

Sterelny and Shanahan may have chosen this strategy because they disliked the hyperpolitical polemics that accompanied the "communist biology" attacks on Gould's writings, and may have decided that a no-politics approach would be a superior alternative. But ignoring the political dimensions of a controversy, when at least one of its participants has himself emphasized the relevance of politics to his scientific views, is not a better approach than distorting the connection between politics and science by sensationalizing it. As a matter of simple fact, Gould's rejection of reductionism cannot be understood outside the context of his political values. Nor can the rest of his scientific opinions be evaluated without reference to the political opinions with which they are intertwined.

Michael Ruse made a good first step toward integrating the discussion of Gould's science with a discussion of his politics in *Mystery of Mysteries*.[78] But Ruse's book is about evolutionary biology in general, and it allots just one chapter to Gould. Gould's particular combination of scientific and political thinking deserves a full-length treatment.

Any time someone such as Ruse, or myself, discusses the combination of science and politics, we, of course, face a question about morality. If, as seems obvious, politics is always about differing conceptions of right and wrong, what becomes of the scientific ideal of objectivity? Not surprisingly, Gould himself addressed this question on several occasions.

Is vs. Ought

Ever since David Hume in the eighteenth century, scientists and philosophers have recognized that there is a difference between stating a descriptive fact about the world and making an evaluative judgment about it. To deduce a moral precept from a presumed fact is to break "Hume's law" or commit the "naturalistic fallacy," and is not acceptable discourse.[79] Yet scientists, philosophers, and ordinary citizens also talk and behave as if there were a connection between what is natural and what is good—who, after all, would recommend that public policy be contrary to human nature? In the early twentieth century, many philosophers of science attempted to fashion clear rules to demarcate the two realms of "is" and "ought," but as the century wore on these attempts came increasingly into disrepute.[80] Everyone agrees that there are some instances of what philosopher John R. Searle terms "brute facts"[81] completely independent of morality and social setting. Even Levins and Lewontin take the time to tell us, "Of course the speed of light is the same under socialism and capitalism."[82] In general, however, to contemporary philosophers the line between is and ought is vague and wavering.

To Gould, however, the line was bright and thick. "Those who recruit Darwin to support a particular moral or political line should remember that, at best," he wrote,

> evolutionary biology may give us some insight into the anthropology of morals—why some (or most) people practice certain values, perhaps for their Darwinian advantage. But science can never decide the morality of morals.[83]
>
> We must remember—and an intellectual's most persistent and nagging responsibility lies in making this simple point over and over again, however noxious and bothersome we render ourselves thereby—that truth and desire, fact and comfort, have no necessary, or even preferred, correlation.[84]

Gould's insistence on the enforcement of Hume's law was rooted in his historical awareness of the way evolution's alleged "facts" were misused by the Social Darwinists of the nineteenth and early twentieth centuries, especially by Herbert Spencer and his epigones.[85] Spencer celebrated the wealth and power of the richest and most powerful as the fulfillment of nature's way, writing that the misery of the poor was evolutionary just deserts. He opposed all efforts on behalf of government to help those on the bottom rung of the social ladder, arguing that public education, the postal service, the regulation of housing conditions, and public health projects were foolish, sentimental, and counter-productive because they prevented society's "unhealthy, imbecile, slow, vacillating, faithless" citizens from being purged, to the benefit of the rest—in other words, because they prevented the elimination of the unfit.[86] Spencer and his American disciple William Graham Sumner were quoted, subsidized, and publicized by captains of industry, who found the idea that their own ascendancy was in tune with nature's way to be intensely agreeable.

Social Darwinism, as an historical phenomenon, was thus a right-wing intellectual movement, and directly contrary to Gould's political values. His position on Spencer's influence was nicely calibrated to give the movement its due, but no more:

> Of course I do not believe that the claims of Social Darwinism directly caused the ills of unrestrained industrial capitalism and the suppression of workers' rights. I know that most of these Spencerian lines functioned as mere window dressing for social forces well in place, and largely unmovable by any academic argument.
>
> On the other hand, academic arguments should not be regarded as entirely impotent, either—for why else would people in power invoke such claims so forcefully?[87]

Nor are these nefarious academic arguments dead. Gould quoted an article in the conservative magazine *National Review* that reviewed the findings of modern sociobiology and concluded that "a Darwinian politics is a largely conservative politics," which teaches us that human nature is based upon self-interest, sexual differences, and natural inequality.[88] To Gould, the conflation of is and ought thus inevitably lent support to social forces that were seeking to justify elite rule. In his mind, Social Darwinism—and by extension the naturalistic fallacy—were by definition right-wing. His invocation of Hume's law was thus part of a strategy to free politics from the burden of the (always conservative) misuse of evolutionary theory.

But Gould was not consistent, for he himself could not resist inferring political lessons from the facts of evolutionary biology. Among a variety of examples, perhaps the best is his use of Lewontin's work on genetic variability to infer basic human equality.

Equality is probably the most important value of the political left. Conversely, the insistence that inequality, whether biological or social, is an ineradicable fact of human life is a central claim of the political right. It is useful for right-wing theorists (normally called conservatives in the United States) if the current understanding of human biology seems to support irreducible inequality. It is useful for left-wing theorists (who may call themselves liberals or progressives) if that same biology seems to diminish the grounds for concluding that groups of humans are significantly different at the genetic level. Gould's position in this biological-ideological continuum was unmistakable. On the subject of racial differences, for example, he reported, "The history of Western views on race is a tale of denial—a long series of . . . retreats from initial claims for strict separation and ranking by intrinsic worth toward an admission of the trivial differences revealed by our contingent history."[89]

How did we know that such racial differences were trivial? (It is useful to notice, in passing, that the word "trivial" is rhetorically loaded. It means not only *small* but *unimportant*, that is, *insignificant.*) We knew because Lewontin's research had revealed that there was no such thing as a "'race gene'—that is, a gene present in all members of one group and none of another."[90] Moreover, "the great preponderance of human variation [at least 80 percent] occurs within groups, not in the differences between them," and therefore,

> Human groups do vary strikingly in a few highly visible characters (skin color, hair form)—and these external differences may fool us into thinking that overall divergence must be great. But we now know that our usual metaphor of superficiality—skin deep—is literally accurate."[91]

In case the reader was tempted to miss the moral, Gould fashioned it into his title for this essay: "Human Equality Is a Contingent Fact of History."[92] The facts of biology refute those who would make political arguments about the superiority of one race or the inferiority of another. Racism is not morally justified. The "is" of biology has compelled a thunderous "ought" in the world of social thought.

Even for a nonbiologist, it is easy to poke holes in this line of reasoning. Eighty percent of genetic overlap among the world's peoples is, in fact, not a particularly impressive proportion, given the tens of thousands of chromosomes on the genome. Moreover, the crucial 20 percent of nonshared genetic material might, to a racist, consist of all the variation that matters. Eighty percent of shared genetics might very well cover the bones, the digestive system, the lymphatic system, and all the rest of the fundamentals of human structure that racists don't care about. Within the varying 20 percent might reside the sorts of differences that racists do care about—intelligence, for example. To infer social equality from a vague report of genetic similarity would thus appear to be going way beyond what the evidence justifies.

More importantly, this and other, similar arguments by Gould must compel us to reevaluate the definition of Social Darwinism. Despite its undoubted historical association with right-wing thinking, the phenomenon is not necessarily conservative in orientation. It appears that progressives, also, can and do infer political conclusions from an interpretation of the facts of biology. Gould was a Social Darwinist, albeit of the left.

It is not a new observation that Darwin's ideas can be used to buttress contradictory political arguments. Half a century ago, Gertrude Himmelfarb summarized her survey of Darwin's influence over social thought by writing:

> In the spectrum of opinion that went under the name of social Darwinism almost every variety of belief was included. . . . It was appealed to by nationalists as an argument for a strong state, and by the proponents of laissez-faire as an argument for a weak state. . . . Militarists found in it the sanction of war and conquest, while pacifists saw the power of physical force transmuted into the power of intellectual and moral persuasion. . . . Political theorists read it as an assertion of the need for inequality in the social order corresponding to the inequality in nature, or alternatively as an egalitarian tract.[93]

Gould, writing after Himmelfarb's book appeared, nevertheless fit into the group she characterized as drawing upon Darwin to produce "egalitarian tracts." Yet he denied it often, even turning the proscription against inferring ought from is into one of his most repeated themes. When encountering such an apparently obtuse inability to recognize theoretic incoherence in one of his own essays in intellectual history—on Leonardo da Vinci, say, or Louis Agassiz, or Charles Darwin—Gould himself was not so much critical as curious, speculating on the origins of the mental blinders that prevented the thinker from thinking critically about himself. In Gould's case, it seems to me that the insistence on the condemnation of all Social Darwinism while practicing a left-wing version of it fulfilled both psychological and political needs. The blanket principle that "nature contains no moral messages" was extremely useful in criticizing right-wing Social Darwinists, whether contemporary or historical.[94] Meanwhile, the failure to recognize his own employment of a left-wing version of the method allowed Gould to construct the positive political arguments he needed to make to his mass audience.

It is my own position that the effort among scientists and philosophers to keep is and ought separate is now and has always been futile. Thinking of an argument in a formal way, as logicians do, the statement of an interpretive conclusion about biological reality functions as a major premise. That is, the statement, "Different groups of human beings differ only insignificantly in their genetic makeup" can be the factual assumption (true or not) from which reasoning starts. All that is necessary is to add what logicians call a "warrant" or "auxiliary assumption" as a minor premise to make the statement into a legitimate moral argument.[95] Such an auxiliary assumption might be, "Only if human beings differ significantly in their genetic makeup is social inequality justified." Given such an explicitly stated assumption, it would then be perfectly acceptable to proceed to the conclusion Gould wants to reach: because human beings do not differ significantly in their genetic makeup, social inequality is not justified. In practice, as mentioned, human beings reason this way all the time. We all want our social arrangements to be consistent with nature, and we are going to argue about whether they are natural regardless of what philosophers and biologists say. By staking an opposing position that was hopelessly self-contradictory, Gould may have made himself a better political advocate, but he turned himself into a worse social theorist.

Science vs. Humanism

Gould was not the only biologist who wanted to derive meaning from the facts of biology; in fact, he may have been typical. George Gaylord Simpson, Theodosius Dobzhansky, and Jacques Monod are only three examples of prominent natural scientists who wrote books toward the ends of their careers in which they took positions on various philosophical issues and drew out some meaning from evolutionary theory.[96] Even Darwin, in a celebrated couple of paragraphs concluding *The Origin*, told his readers that the process of evolution had "ennobled" each species because it meant that all creatures "will tend toward perfection."[97]

Nevertheless, there is also a strong tradition in science of being suspicious of the power of humanism to mislead. Many scientists would agree with Richard Dawkins's complaints about the "power of poetic imagery to inspire bad science." In particular, Dawkins accused Gould of being a "bad poet," and he put a good deal of energy into criticizing not just Gould's ideas but his whole style of exposition.[98]

Not surprisingly, however, Gould embraced the project of combining the truths of science with the interpretations of humanism. His view of the fusion was the opposite of Dawkins; he believed that in the right hands it illuminated rather than obscured. He told us often, "I am a humanist at heart, and I love, best of all, the sensitive and intelligent conjunction of art and nature."[99] Indeed, his very frequent discussions of music, architecture, philosophy, literature, and history in conjunction with nature provided all the evidence a reader needed to conclude that his heart was as he claimed. With Gould, however, although the musings on culture had intrinsic value, they also had purpose. Humanism was a pleasure in itself, but it was also a means to an end, which was ultimately political.

As Gould told us in 2000, he did not begin writing his popular essays in 1973 with his strategy already conceived. He thought then that "I would use my humanistic and historical interests as a 'user-friendly' bridge to bring my readers into the accessible world of science." But,

> Over the years . . . this mere device . . . became an explicit centrality. . . . I finally realized that . . . nearly five hundred years of tradition had established and validated . . . the essay as a genre dedicated to personal musing and experience, used as a gracious entrée, or at least an intriguing hook, for discussion of general and universal issues.[100]

In nearly three decades of popular-science writing, Gould covered a large number of general and universal issues. There were a number, however, to which he returned repeatedly. I am going to examine one in particular because he discussed it so often that we can conclude it was one of, if not the most, important theme in his writing.

As I will show in the coming chapters, Gould frequently attempted to persuade his readers, lay and professional, to accept two grand conclusions about the history of life. The first was that evolution is nondirectional; it does not point toward any types of organisms. We may not conclude, for example, that intelligence evolved because intelligent creatures are in some way superior. The second, which is closely allied to the first, was that human beings are in no sense special. From a biological perspective, humans are no more highly evolved, and certainly not morally superior to, bacteria. In future chapters I will examine the controversy within biology over these two claims. Here I will indicate the political purpose underlying their importance in Gould's thought.

The two principles of nondirectionality and human nonspecialness work in unison toward an ideological goal. That goal is best understood in relation to a favorite quotation from Sigmund Freud. Gould had read a great deal of

Freud, and referred to many of his arguments while making his own points. One of Freud's statements in particular, however, occupied a central place in Gould's pantheon of insights. He quoted it early in his first book, and, by my count, reproduced it in two more books and summarized or alluded to it in seven others:

> Humanity has in course of time had to endure from the hands of science two great outrages upon its naive self-love. The first was when it realized that our earth was not the center of the universe, but only a speck in a world-system of a magnitude hardly conceivable. . . . The second was when biological research robbed man of his particular privilege of having been specially created and relegated him to a descent from the animal world.[101]

"Freud claimed," in a comment Gould repeated many times in the essays, whether or not he referred to the Austrian psychiatrist's specific words, "that all important scientific revolutions share the ironic property of deposing humans from one pedestal after another of previous self-assurance about our exalted cosmic status."[102] Gould returned to the point so often because he wished to endorse it. Yes, that is exactly what scientific enlightenment does: it removes the props from human hubris. Or, at least, that would be the tendency of science, if the public truly understood and accepted it.

Unfortunately, however, biology has not been as thorough as it should have been in the reeducation project. "Darwin removed this keystone of false comfort [human cosmic significance] more than a century ago, but many people still believe that they cannot navigate our earthly vale of tears without such a crutch."[103] People continue to resist their existential demotion; we "grasp at the straw of progress (a desiccated ideological twig) because we are still not ready for the Darwinian revolution."[104] Therefore, "Darwin's revolution remains incomplete to this day because we spin-doctor the results of evolution to preserve our pedestal of arrogance by misreading the process as a predictable accumulation of improvements, leading sensibly to the late appearance of human intelligence as a culmination."[105]

The essential question then becomes, why does it matter? Supposing that Gould is correct, and humans continue to resist the Darwinian solvent that should otherwise be dissolving their self-esteem as a species—so what? What harm does it do if humans think that nature has awarded them the title of most evolved?

Gould's answer to this question revealed both his central moral assumption and the way he wove morality, science, and politics into an ideological whole. It matters, he said, because "success in our profession's common battle for preserving biodiversity requires the reorientation of human attitudes toward other species—from little care and maximal exploitation to interest, love, and respect. How can this change occur if we continue to view ourselves as better than all others by cosmic design?"[106] Humans are the most arrogant and rapacious species on the planet, and the arrogance justifies the rapacity. If we are to save the other species from the human-caused holocaust, we must reeducate humans out of their arrogance. In this sense, it would perhaps be accurate to describe Gould as an antihumanist humanist. One of his important goals in bringing in extrascientific material to his essays is to render his attack on human self-esteem more palatable to its intended audience.

When Gould thus speaks of "our profession's common battle," he is not mischaracterizing his colleagues. In his campaign to educate, embarrass, or frighten humanity into pulling back from biocide, he was only one general in a war that had become desperately important to evolutionary biologists in general. Although Dawkins and others disliked Gould's "bad poetry," many prominent biologists implicitly endorsed his attempt to propagandize on behalf of the diversity of life. Take, for example, Edward O. Wilson, who, as we will see, engaged in long-running battles with Gould on other scientific topics. Wilson is one of the many biologists who have published eloquent pleas for people to stop their assault on the planet's web of life:

> Signals abound that the loss of life's diversity endangers not just the body but the spirit. If that much is true, the changes occurring now will visit harm on all generations to come. The ethical imperative should therefore be, first of all, prudence. We should judge every scrap of biodiversity as priceless while we learn to use it and come to understand its meaning to humanity.[107]
>
> The political process in American democracy, with rare exceptions, does not start at the top and work its way down to the voting masses. It proceeds in the opposite direction. . . . An alliance between science and religion . . . may be the only way to protect life on earth, including, in the end, our own.[108]

On this, the most important political issue of their profession, Gould and Wilson had no disagreements. As with Wilson, all of Gould's undoubted competence as a scientist was in league with his skill as a rhetorician in the service of this crucial collective battle for public opinion. The very way he defined and explained his craft was a part of a strategy that aimed beyond science to humanistic communication, and thence to political influence.

notes

1. Steven Weinberg, *Facing Up: Science and Its Cultural Adversaries* (Cambridge, MA: Harvard University Press, 2001), p. 84.

2. Peter Godfrey-Smith, *Theory and Reality: An Introduction to the Philosophy of Science* (Chicago: University of Chicago Press, 2003), pp. 58–66, 71–74.

3. John A. Moore, *Science as a Way of Knowing: The Foundations of Modern Biology* (Cambridge, MA: Harvard University Press, 1993), p. 503.

4. For discussions of these issues, see Godfrey-Smith, *Theory and Reality*, pp. 8, 23, 39, 43, 126–28, 131, 134; John R. Searle, *The Construction of Social Reality* (New York: Simon and Schuster, 1995), pp. 150–58, 163–67, 178–79, 183, 196, 197; Philip Kitcher, *The Advancement of Science: Science Without Legend. Objectivity Without Illusions* (New York: Oxford University Press, 1993), pp. 94, 138, 170.

5. Stephen Jay Gould, *The Structure of Evolutionary Theory* (Cambridge, MA: Harvard University Press, 2002), p. 44.

6. Nancy Thorndike Greenspan, *The End of the Certain World: The Life and Science of Max Born* (New York: Basic Books, 2005).

7. Godfrey-Smith, *Theory and Reality*, p. 98.

8. Thomas Kuhn, *The Structure of Scientific Revolutions*, 3rd. ed. (Chicago: University of Chicago Press, 1996), pp. 150, 170.

9. Ibid., pp. 66–110.

10. Ibid., pp. 150, 170.

11. David L. Hull, *Science as a Process: An Evolutionary Account of the Social and Conceptual Development of Science* (Chicago: University of Chicago Press, 1988), pp. 1, 11.

12. Searle, *Social Reality*, p. 183.

13. Gould, *Structure*, p. 967.

14. Stephen Jay Gould, *Time's Arrow, Time's Cycle* (Cambridge, MA: Harvard University Press, 1987), p. 7.

15. The term is Hull's: *Science as a Process*, pp. 1–4, 6, 10–11.

16. Karl Popper, *Conjectures and Refutations: The Growth of Scientific Knowledge* (New York: Harper and Row, 1965), p. 226.

17. Ibid., p. 220.

18. Stephen Jay Gould, *Eight Little Piggies: Reflections in Natural History* (New York: W. W. Norton, 1993), p. 430.

19. Stephen Jay Gould, *The Flamingo's Smile: Reflections in Natural History* (New York: W. W. Norton, 1985), pp. 417–18.

20. Stephen Jay Gould, *Leonardo's Mountain of Clams and the Diet of Worms* (New York: Three Rivers Press, 1998), p. 155.

21. Stephen Jay Gould and Niles Eldredge, "Punctuated Equilibria: The Tempo and Mode of Evolution Reconsidered," *Paleobiology* 3 (1977): 115–51; Gould and Eldredge, "Punctuated Equilibrium Comes of Age," *Nature* 366 (1993): 223–27.

22. Gould, *Eight Little Piggies*, p. 412.

23. Stephen Jay Gould, *Ever Since Darwin: Reflections in Natural History* (New York: W. W. Norton, 1977), p. 12; Gould, *Structure*, p. 122.

24. Marjorie Grene, "Perception, Interpretation, and the Sciences: Toward a New Philosophy of Science," in David J. Depew and Bruce H. Weber, eds., *Evolution at a Crossroads: The New Biology and the New Philosophy of Science* (Cambridge, MA: MIT Press, 1985), pp. 6–7.

25. Matthew Mohan and Bernard Linsky, "Introduction," in Mohan and Linsky, eds., *Philosophy and Biology* (Calgary: University of Calgary Press, 1988), p. 2.

26. Carl Hempel, *Philosophy of Natural Science* (Englewood Cliffs, NJ: Prentice-Hall, 1966), p. 51; Philip Gasper, "Causation and Explanation," in Richard Boyd, Philip Gasper, and J. D. Trout, eds., *The Philosophy of Science* (Cambridge, MA: MIT Press, 1991), pp. 289–92.

27. James Bohman, *New Philosophy of Social Science: Problems of Indeterminacy* (Cambridge, MA: MIT Press, 1991), pp. 18–30.

28. Mohan and Linsky, "Introduction;" Elliott Sober, *The Nature of Selection* (Cambridge, MA: MIT Press, 1984); Alex Rosenberg, *The Structure of Biological Science* (Cambridge: Cambridge University Press, 1985); Michael Ruse, *Taking Darwin Seriously: A Naturalistic Approach to Philosophy*, 2nd ed. (Amherst, NY: Prometheus Books, 1998); Daniel C. Dennett, *Darwin's Dangerous Idea: Evolution and the Meanings of Life* (New York: Simon and Schuster, 1995); Ernst Mayr, *Toward a New Philosophy of Biology: Observations of an Evolutionist* (Cambridge, MA: Harvard University Press, 1988); Moore, *Science as a Way of Knowing*.

29. Stephen Jay Gould, *Wonderful Life: The Burgess Shale and the Nature of History* (New York: W. W. Norton, 1989), p. 279.

30. Stephen Jay Gould, *An Urchin in the Storm: Essays about Books and Ideas* (New York: W. W. Norton, 1987), p. 78.

31. Gould, *Structure*, pp. 1183, 1205.

32. Gould, *Wonderful Life*, p. 281.

33. Stephen Jay Gould, *The Hedgehog, the Fox, and the Magister's Pox* (New York: Three Rivers Press, 2003), p. 227.

34. Stephen Jay Gould, *Hen's Teeth and Horses Toes: Further Reflections in Natural History* (New York: W. W. Norton, 1983), p. 123.

35. Gould, *Wonderful Life*, p. 279.

36. Gould, *Flamingo's Smile*, p. 111.

37. Gould, *Hen's Teeth*, p. 256.

38. Gould, *Urchin*, p. 184.

39. Gould, *Structure*, p. 1332.

40. Stephen Jay Gould, *Dinosaur in a Haystack: Reflections in Natural History* (New York: Harmony Books, 1995), p. 409.

41. Ibid.

42. Gould, *Magister's Pox*, p. 227.

43. Gould, *Structure*, p. 1332.

44. George Gaylord Simpson, *This View of Life: The World of an Evolutionist* (New York: Harcourt, Brace, and World, 1964), p. 128.

45. Gould, *Structure*, pp. 761, 803, 1028.

46. Antoni Hoffman, *Arguments on Evolution: A Paleontologist's Perspective* (New York: Oxford University Press, 1989), pp. 43–44.

47. Jefrey S. Levinton, *Genetics, Paleontology, and Macroevolution* (Cambridge: Cambridge University Press, 2001), p. 374.

48. Moore, *A Way of Knowing*, p. 219.

49. Mark A. S. McMenamin, "The Origins and Radiation of the Early Metazoa," in K. C. Allen and D. E. G. Briggs, eds., *Evolution and the Fossil Record* (Washington, DC: Smithsonian Institution, 1990), pp. 88–94; John Maynard Smith and Eors Szathmary, *The Origins of Life: From the Birth of Life to the Origin of Language* (Oxford: Oxford University Press, 1999), pp. 109–10.

50. Richard Levins and Richard Lewontin, *The Dialectical Biologist* (Cambridge, MA: Harvard University Press, 1985), pp. 1, 3.

51. Ibid., p. 271.

52. Ullica Segerstrale, *Defenders of the Truth: The Sociobiology Debate* (Oxford: Oxford University Press, 2000), pp. 41–42, 102–105, 203–205.

53. Richard C. Lewontin, Steven Rose, and Leon J. Kamin, *Not in Our Genes: Biology, Ideology, and Human Nature* (New York: Pantheon Books, 1984), p. 9.

54. Levins and Lewontin, *Dialectical Biologist*, p. 5.

55. Ibid., p. 273.

56. Weinberg, *Facing Up*, p. 118.

57. Ibid., p. 7.

58. Ibid., p. 68.

59. Ibid., p. 290.

60. Jacques Monod, *Chance and Necessity: An Essay on the Natural Philosophy of Modern Biology* (New York: Vintage, 1972), p. 79.

61. Peter Singer, *A Darwinian Left: Politics, Evolution and Cooperation* (New Haven, CT: Yale University Press, 2000), p. 29.

62. Ernst Mayr, *Toward a New Philosophy of Biology: Observations of an Evolutionist* (Cambridge, MA: Harvard University Press, 1988), p. 1; see also pp. 10–11, 14, 21, 26, 405, 449.

63. Gould, *Urchin*, p. 151.

64. Stephen Jay Gould, "Fulfilling the Spandrels of World and Mind," in Jack Selzer, ed., *Understanding Scientific Prose* (Madison: University of Wisconsin Press, 1993), pp. 319–20.

65. Richard Dawkins, *The Selfish Gene* (Oxford: Oxford University Press, 1976), p. 2.

66. Ibid., p. 243.

67. Richard Dawkins, *The Extended Phenotype: The Long Reach of the Gene* (Oxford: Oxford University Press, 1982), pp. 222–27.

68. Philip Kitcher, "1953 and All That: A Tale of Two Sciences," pp. 553–70, and Elliott Sober and Richard Lewontin, "Artifact, Cause, and Genic Selection" pp. 571–88, in Boyd, Gasper, and Trout, *Philosophy of Science*; Hull, *Science as a Process*, pp. 408–30.

69. Gould, *Structure*, p. 618.

70. Ibid., p. 632.

71. Richard Dawkins, *The Blind Watchmaker: Why the Evidence of Evolution Reveals a Universe without Design* (New York: W. W. Norton, 1996), p. xiii.

72. Gould, *Wonderful Life*, p. 323; Gould, *Structure*, p. 936.

73. Gould, *Eight Little Piggies*, p. 309.

74. Ibid., pp. 310–11.

75. Kitcher, "1953 and All That," p. 567.

76. Kim Sterelny, *Dawkins vs. Gould: Survival of the Fittest* (Cambridge: Totem Books, 2001).

77. Timothy Shanahan, *The Evolution of Darwinism: Selection, Adaptation, and Progress in Evolutionary Biology* (Cambridge: Cambridge University Press, 2004), pp. 63–70, 137–42, 148–51, 288–93.

78. Michael Ruse, *Mystery of Mysteries: Is Evolution a Social Construction?* (Cambridge, MA: Harvard University Press, 1999), pp. 135–52.

79. Michael Ruse, "Evolutionary Ethics: Healthy Prospect or Lost Infirmity?" in Mohan and Linsky, *Philosophy and Biology*, p. 58.

80. Hillary Putnam, *The Collapse of the Fact/Value Dichotomy and Other*

Essays (Cambridge, MA: Harvard University Press, 2002), pp. 7–45; Grene, "Perception," p. 2.

81. Searle, *Social Reality*, pp. 2, 56.

82. Levins and Lewontin, *Dialectical Biologist*, p. 4.

83. Stephen Jay Gould, *I Have Landed: The End of a Beginning in Natural History* (New York: Three Rivers Press, 2003), p. 221.

84. Stephen Jay Gould, *Bully for Brontosaurus: Reflections in Natural History* (New York: W. W. Norton, 1991), p. 57.

85. I discuss this topic in more detail in *The Paradox of Democratic Capitalism: Politics and Economics in American Thought* (Baltimore: Johns Hopkins University Press, 2006), pp. 107–11, 116, 124.

86. Spencer, quoted in Stephen Jay Gould, *The Lying Stones of Marrakech: Penultimate Reflections in Natural History* (New York: Harmony Books, 2000), p. 261.

87. Ibid., p. 264.

88. Gould, *I Have Landed*, p. 220.

89. Gould, *Flamingo's Smile*, p. 187.

90. Ibid., p. 196.

91. Ibid.

92. Ibid., p. 185.

93. Gertrude Himmelfarb, *Darwin and the Darwinian Revolution* (Chicago: Ivan R. Dee, 1996), p. 431.

94. Gould, *Hen's Teeth*, p. 43.

95. Stephen Toulmin, *The Uses of Argument* (Cambridge: Cambridge University Press, 1958), pp. 98, 125.

96. George Gaylord Simpson, *This View of Life: The World of an Evolutionist* (New York: Harcourt, Brace, and World, 1964); Theodosius Dobzhansky, *The Biological Basis of Human Freedom* (New York: Columbia University Press, 1956); Monod, *Chance and Necessity*.

97. Charles Darwin, *On the Origin of Species* (New York: Barnes and Noble, 1859, 2004), p. 384.

98. Richard Dawkins, *Unweaving the Rainbow: Science, Delusion, and the Appetite for Wonder* (Boston: Houghton, Mifflin, 1998), pp. 180, 193–203.

99. Gould, *Diet of Worms*, p. 2.

100. Gould, *Lying Stones of Marrakech*, p. 2.

101. Gould, *Ever Since Darwin*, pp. 16–17.

102. Gould, *Diet of Worms*, p. 286.

103. Gould, *I Have Landed*, p. 217.

104. Stephen Jay Gould, *Full House: The Spread of Excellence from Plato to Darwin* (New York: Three Rivers Press, 1996), p. 29.

105. Gould, *Diet of Worms*, p. 286.

106. Gould, *Full House*, p. 27.

107. Edward O. Wilson, *The Diversity of Life* (Cambridge, MA: Harvard University Press, 1992), p. 351.

108. Edward O. Wilson, "Apocalypse Now: A Scientist's Plea for Christian Environmentalism," *New Republic*, September 4, 2006, p. 18.

Chapter 3

the contours of history

For some time, the profession of history has been in turmoil. Ever since postmodern approaches to truth, which, in essence, deny that truth exists, began to seep into historical thought at the beginning of the twentieth century, the practice of historical scholarship has been experiencing what J. Daryle Charles called an "extended epistemological crisis."[1] All historians agree that there must be, in Leonard Krieger's words, "some kind of coherence, whether explanatory or generalizing" to lend intelligibility to otherwise chaotic facts, but there is no consensus on what type of conceptual framework to endorse.[2] Whether any given historian embraces Marxism, or feminism, or poststructuralism, or rationalism, or idealism, or empiricism, or any one of the other approaches to the discipline, he or she is bound to be in the minority.[3]

Nevertheless, there is one aspect of history about which all historians agree, something that is so basic it is not even explicitly addressed in historical writing: history is about humans. As Krieger reports, historians study "what some men have done in the past . . . men as such do all the time with reference to the past" and "what historians do with the past."[4] It would not occur to Krieger, or most other historians, to conceptualize "history" as the entire record of life on earth.

That was, however, exactly how Stephen Jay Gould characterized history. He was, he told us, "a historian at heart."[5] But to Gould, it would have been nonsensical to separate human history from its evolutionary context. The sort of history practiced by Charles, Krieger, and Lloyd would have seemed jejune to him, draining humans of the meaning their existence acquires through placement in the natural order. It is exactly because "we are the offspring of history"—that is, of history in its cosmic fullness—that we must understand history in order to understand ourselves.[6]

Despite Gould's hopefulness, in practical terms the professions of history and evolutionary biology are not closely connected in the universe of scholarly discourse. The followers of each discipline are not even housed in the same colleges in most universities, one being in liberal arts, the other in natural science. Yet there is an irony in this separation, for evolutionary biology possesses what the historians wish they had—an organizing conceptual framework common to everyone in the profession. Because all evolutionary biologists share an allegiance to the theoretical framework that derives from Darwin in 1859, they have not had to undergo the extended epistemological crisis that has tormented actual historians for a century.

In general outline, the conceptual framework—the "theory"—is deceptively simple. Within each species, there is always a considerable amount of variation. Some individuals in a species will be larger, or faster, or smarter, or possess better vision or a more tolerant digestive system, or toxic saliva. This is the principle of *ubiquitous variation.* All species produce more offspring than can possibly survive in a dangerous world. This might be termed the principle of *latent potential overpopulation.* On the average and in the long run, those offspring that are better adapted to environmental challenges will survive to breed, while the more weakly adapted will perish, or, more simply, fail to breed. The survivors will thus pass on their better eyesight, or toxic saliva, to their own offspring. This is the principle of *natural selection.* Over deep time, the passing on of small, superior differences from generation to generation has resulted in a continuous set of elaborations that are on display in the luxurious diversity of life in the present world. Darwinians term this historical consequence *descent with modification.* Although Darwin had no knowledge of the mechanism by which variations were passed on, and was completely wrong in his attempts to theorize about it, the arrival of Mendelian genetics in 1900 and the understanding of the structure of DNA in 1953 have filled in the gaps of the theory, resulting in the "Neo-Darwinian Synthesis" that today structures the thinking of all biologists.[7]

Although, as we shall see shortly, there is room for many varied and intense controversies of detail within the basic Darwinian paradigm, it nevertheless provides a grounding of fundamental conceptual agreement to everyone working within the sciences of life. No matter how many disagreements Gould and his allies may generate with other evolutionary biologists, they all share a language, worldview, and epistemology that are beyond the grasp of historians of humanity.

Notwithstanding the paradigmatic commonality of evolutionary biology, its disagreements of detail can be ferocious. As it happened, during his career Gould himself generated many of the most interesting conflicts, and participated in others. From the perspective of 2009, it can be seen that in most of those disagreements he represented a rather small minority of scientists. Nevertheless, many of the conflicts in which he played a part reverberated outside the confines of the science of life, and because Gould was such a superb writer, he was able to convey his side of the conflicts to a large, nonprofessional audience. A major principle of democratic politics is that "the spectators are an integral part of the situation, for, as likely as not, the *audience* determines the outcome of the fight. . . . The outcome of every conflict is determined by the *extent* to which the audience becomes involved in it."[8] Because Gould reached such a large readership, he was always threatening to bring a nonscientific crowd into scientific debate. When his subject was Darwinism vs. creationism, other scientists appreciated the way he could rally the rational public against the barbarians. But when his subject was a conflict interior to evolutionary biology, such as the proper way to conceptualize history, scientists who disagreed with him resented his seeming willingness to involve nonprofessionals in professional disputes.

Although Gould's literary charm partly underlay his ability to reach outside the boundaries of his profession, another part of his success was based on the fact that his subject matter was inherently compelling. The history of life attracted attention from millions of people outside of science because of a fact that few scientists, including Gould, liked to admit publicly. Despite his claim, discussed in the previous chapter, that we could not infer morality from nature, he believed that history contained lessons about human life. Because they understood the moral importance of life's history, and because Gould was so accessible, millions of his readers were available to become part of the audience that might decide internal scientific disagreements.

Furthermore, to Gould, always, those lessons were consistent with his left-wing politics. It would be crude and misleading to say that he inferred his evolutionary biology from his politics. It would be equally deceptive to characterize his politics as being inferred from his biology. An accurate characterization would be that his views on the history of life were seamlessly consistent with his political values. His ideas were all of a piece.

But why should the shape of the history of life contain political lessons? The answer is that the facts, or supposed facts, of evolution suggest implications for the way humans interpret their own place in the cosmos. The most basic of these implications is the one dividing scientists from creationists—the implication of Darwinism is that no supernatural force fashioned life, including human life. It does not take great sophistication, in the early years of the twenty-first century, to understand that this divide marks a political conflict.

But there are also more subtle implications to draw out of differing views of the contours of history. A subtle thinker, Gould was aware of these implications, and the way they could be advanced or retarded by one evolutionary narrative as opposed to another. He was not the only evolutionary biologist to understand the ultimate consequences of varying scientific narratives of life, and many of the attacks on his theories by other scientists are best understood within the context not just of methodological disagreement but of political quarrel. Other scientists were always uneasy with Gould's potential power to bring the audience into political conflict.

Because Gould was a leftist, he endorsed the primary value of *equality*. Whether he was an actual Marxist or a democratic socialist or a modern liberal does not matter in this context. It is characteristic of the left side of the political spectrum that its denizens believe in human moral equality. People to the right of the spectrum are comfortable with the inequalities that exist around them, and they customarily oppose efforts by government to change them. People to the left are offended by inequalities, and they customarily support efforts to change them.

The moral argument between the right and the left over inequality generally comes down to differing estimations of the importance of *merit* versus *chance*. It is characteristic of the political right that its adherents typically argue that those who have more of anything in society have earned their greater possessions through greater merit, effort, and willingness to take risks. To the right, the rich and powerful merit their rewards, and the poor and weak likewise merit theirs. To the right, therefore, there is rarely a good reason for government to engage in a campaign of redistribution. To the left, in contrast, the present distribution of wealth and power is largely a result of happenstance. By and large, rich and powerful people do not deserve their lot in life, and the poor and weak do not deserve theirs. Leftists such as Gould thus consistently support redistribution of whatever there is to get.[9]

It may seem that government redistribution of resources is of little relevance to the history of extinct species in deep time. Certainly there is no direct connection between modern political ideologies and the evolution of nonhuman species over 3.8 billion years. But the theory of evolution is a narrative interpretation of facts, and facts about life always have moral implications, at least to many observers. Accepted patterns of evolution link to political ideologies through the implied connecting concepts of merit and chance.

If the history of life is the record of relatively smooth upward transition from simplicity to complexity and from blind blundering to intelligence, and if the species alive today are in some sense history's winners, then the principle of merit seems to shine forth from the record of the rocks. If, on the other hand, the history of life is one of randomness and luck, with no progression evident, and living species are simply the beneficiaries of chance and necessity, then accident governs the universe and there is no principle of merit evident in life. In particular, if humans are the product of an inexorable tendency to improvement in evolution, then they are, in some sense, the most evolved creatures on earth, and have reason to be smug about it. In contrast, if they are merely the result of a billion lucky breaks, then they are simply the incidental outcome of a cosmic crapshoot, and have no cause to think of themselves as anything special.

Because Gould was a political leftist, he embraced the notion that concepts of progress, and therefore of merit, had no relevance to the history of life. Instead, he consistently argued that evolution was unpredictably contingent at every juncture. When he mused, in *Wonderful Life* in 1989, "Perhaps the Grim Reaper of anatomical designs is only Lady Luck in disguise," he was stating not only his opinion about the history of life but an important tenet of his political philosophy as well.[10] Other scientists who either disagreed with his politics or rejected the implications Gould drew from the patterns of history argued with Gould about his science. The surface substance of most of the controversies in which he engaged consisted of arcane disputes about methodology and the interpretation of evidence. But underlying all the scientific controversies lay a subtext of political dispute, generally implicit, sometimes publicly denied, but ever present.

A Shaky Orthodoxy

Like all pioneers, Darwin was not completely consistent in knitting all the aspects of his theory together, leaving future historians of ideas to argue about what he "really" meant. Furthermore, his theory as published in the first edition was immediately subject to some shrewd criticisms. Although most of these objections have been rendered moot by the advance of biological science in the last century and a half, in his own lifetime Darwin attempted to refute them by modifying his theory in five subsequent editions of *Origins*. His detailed modifications sometimes contradicted the manifest tendency of his theory as a whole. As a result, it is possible for sharp-eyed scholars to find many contradictions in Darwin, and argue that this or that generalization about his theory is falsified by this or that quotation from this or that edition. When anybody tries to summarize a theoretical advance in evolutionary biology by explaining how it differs from Darwin's original formulation, therefore, they are subject to sniping from close readers who can point to a passage or two in which Darwin said the opposite of what the author portrays him as saying. And if this is true for Darwin, it is plurally true for the theorists of the Evolutionary Synthesis in the twentieth century, who were also not completely consistent in their writings over careers that spanned decades.

In this book I will try to summarize the theories of Darwin, the authors of the Neo-Darwinian Synthesis, Gould, and Gould's collaborators and critics by reporting their main points and general conceptual thrust, so that I may explicate and evaluate the evolutionary controversies that are relevant to my project. I will not try to referee every nuance, ambiguity, and inconsistency.

Darwin's theory emphasized slow, tiny alterations over immense spans of time. Although the title of his book promised to explain the origin of species, his theory presented a novel and unorthodox interpretation of what species actually are. The reigning view in Darwin's time, which is the common-sense view for all time, was essentialist; that is, it held that species were discrete particles of reality, each one clearly demarcated from all other species. A horse and a zebra share some similarities, as do a cobra and a rattlesnake, but it is nevertheless easy to sort the individuals in each species into one bin or another, using a variety of classificatory criteria. To Darwin, however, this discreteness of species was an illusion created by the fact that humans behold creatures during a snapshot of time. If people could see life within the fullness of evolution, however, they would understand that species are not discrete things, but a frozen moment of a temporal continuum, constantly in motion from one physical construct—an individual phenotype—to another. Humans intercept organisms that happen to be within the stream of time at their own moment, and interpret the transient individual they see as a representative of a species, a structured type. Instead, they should interpret any given organism as we might today interpret a single frame of a motion picture film stock—as one arrested moment among thousands of arrested moments, no individual of which is any more defining than any other. Not having the motion-picture metaphor available to him, Darwin expressed the idea differently:

> I look at the term *species* as one arbitrarily given for the sake of convenience to a set of individuals closely resembling each other, and that it does not essentially differ from the term *variety*, which is given to less distinct and more fluctuating forms. . . . It may be said that natural selection is daily and hourly scrutinizing, throughout the world, every variation, even the slightest; rejecting that which is bad, preserving and adding up all that is good; silently and insensibly working . . . at the improvement of each organic being. . . . We see nothing of these slow changes in progress, until the hand of time has marked the long lapse of ages.[11]

Richard Dawkins, the best modern spokesman for orthodox Darwinism, has characterized the views of a large proportion of his colleagues by writing that "taking a long view of the entirety of human evolutionary history [we] cannot see 'the species' as a discrete entity at all. [We] can see only a smeary continuum."[12]

If species graded into one another very slowly, one tiny change at a time, then it followed that the history of life must be one of an extraordinary number of intermediate species. Indeed, every species was intermediate, unless the world ended during its lifetime. Darwin clearly saw what this derivation from his theory must mean for the fossil record: "If my theory be true, numberless intermediate varieties, linking most closely all the species of the same group together, must assuredly have existed."[13]

Darwin was aware even as he wrote that the fossil record did not provide much support for the expectations raised by his theory. Rather than showing an insensible gradation of one species into another, the layers of rocks that had been investigated up to the 1850s displayed species that, in general, seemed to arrive suddenly ("suddenly" being a relative term that, to paleontologists, might mean millions of years), persisted for a short or long duration relatively unchanged, then disappeared abruptly. He explained the apparently uncooperative evidence as being caused by the extremely fragmentary nature of the fossils that had, until 1859, been discovered. He also expressed hope that with the advance of scientific knowledge in the future, the gaps might be filled in and at least some of the intermediate sequences mapped. When, two years after the appearance of *Origin*, a fossil of an *Archaeopteryx*, a seemingly transitional organism between reptiles and birds, was discovered in a quarry in Germany, Darwinists concluded that the theory had been confirmed.[14]

Yet from 1859 until the 1970s there was always an undercurrent of dissent about the adequacy of Darwin's theory to explain the pattern of life's history. Christian creationists, of course, always rejected the theory. No number of transitional fossils would have satisfied their demand for more evidence. More importantly, within the confines of evolutionary biology there was a continuing thread of doubt that a theory emphasizing slow, persistent change over hundreds of millions of years could account for the evolutionary record, either theoretically or empirically.

When the Neo-Darwinian Synthesis came together during the 1940s, the surface unanimity of outlook covered enduring differences of emphasis. Those scientists whose specialty was statistical modeling or genetics, such as R. L. Fisher and Theodosius Dobzhansky, tended to embrace the view of evolution as consisting of a multitude of small steps.[15] The scientists who worked more with whole organisms, whether living or extinct, such as George Gaylord Simpson and Ernst Mayr, tended to be less convinced that transformations of one species into another had to take place at rates of geologic time.[16] But Simpson, stung by a hostile reaction to his work by other members of the Synthesis, had recanted on his theory of rapid "quantum evolution" by 1953 and had joined the gradualist consensus.[17] And Mayr complained in 1988 that his own forays into nongradualist theory during the 1950s had been "almost universally ignored."[18]

The Darwinist/Synthetic emphasis on a long series of slow steps contained an important implication. If everything that happened in evolution took place as the result of a sequence of minute mutations, then the whole pattern of historical change since the first molecule replicated itself in the primordial soup roughly 3.8 billion years ago could be explained with reference to alterations at the level of the individual organism. Individuals of the same species had, over eons, given rise to differing species, which, as their lineages diverged, accumulated changes and gradually gave rise to different genera, families, orders, classes, phyla, and kingdoms. The contours of *macroevolution*—large-scale patterns evident in the history of life— were caused simply by the forces at work in *microevolution*—the slow accumulation of tiny changes within individuals and species. There was no need for a separate theory of macroevolution; learning about stability and change in genetics would fulfill all the needs of science to understand the history of life.

The Pattern in the Rocks

Doing field research for their dissertations in paleontology during the late 1960s, both Gould and Niles Eldredge began with orthodox expectations. Eldredge, studying trilobites, and Gould, studying land snails, anticipated their ability to document the long, slow transformation of one species into another. Over the course of their researches, they both became frustrated with the lack of evidence they seemed to be examining. Eldredge later wrote that all his trilobite species seemed to jump full-blown into the fossil record, and as for documenting slow transitions, "I was not prepared for the inertial stability of my fossils. They didn't seem to *want* to change."[19]

At some point about 1970 Eldredge made a conceptual leap and began to consider the possibility that the fossil record, far from misleading through its incompleteness, actually provided a good documentation of life's history. He began to think seriously about a new way to conceptualize patterns of evolution. Perhaps species tended to maintain a consistent body plan for long stretches of time, but then when subjected to severe environmental stress, or when some individuals were geographically isolated from others of their species (on an island, say), they might undergo fairly rapid changes, resulting in new species. He had already published an article in the professional journal *Evolution* in which he had reported that his trilobite samples appeared to exhibit a great deal of "morphological stability," that new species materialized in the rock strata with surprising suddenness, and that the traditional way of conceptualizing evolution as slow and steady did not seem to fit the facts.[20] Thinking about these ideas together, he decided that the time had come to challenge the dominant paradigm of evolution among paleontologists.

Eldredge had been discussing his ideas with Gould as they developed.

Gould was receptive to the new way of looking at the pattern of life's history, and gave the alternate theoretical model a name: *punctuated equilibrium*. In 1972, and in subsequent publications, the two former fellow graduate students propounded their theory of evolutionary change, a theory that made them both famous and infamous, and ensured their lasting impact within natural science and the larger world.[21]

The theory of punctuated equilibrium (or equilibria) was an attempt to suggest a new view of the *pace* of evolution as envisioned by orthodox Darwinism, a new *mode* of change as expressed in its explanation of the way in which small changes in individuals and populations translate into transformations of species, a new interpretation of the *evidence* paleontologists had traditionally adduced to support the theory, and therefore, a new *conceptual framework* to use to understand the story of life. Eldredge and Gould emphasized that "we postulate no 'new' type of selection."[22] They continued to endorse natural selection and descent with modification as the master ideas in historical explanation. But they provided a radical alternative view of the way that natural selection functioned.

The main tenet of orthodox Darwinism, they argued, was *phyletic gradualism*, that is, "a long and insensibly graded chain of intermediate forms" linking one species to another over deep time.[23] Their dissatisfaction with this orthodoxy was based on the perception that, after more than a century of industrious digging by paleontologists, it was still supported by "blatantly inadequate data."[24] Darwinian orthodoxy led everyone to expect that the earth's rock layers must contain the preserved evidence of a series of smoothly transitional sequences of types. Instead, as Gould later asserted in one of his popular essays, "the extreme rarity of transitional forms in the fossil record persists as the trade secret of paleontology."[25] Rather than the pattern that orthodox theory predicts, scientists beheld in the fossil record a pattern marked by the sudden appearance of all or most species, which then persisted relatively unchanged for their entire existence, and at some point became extinct. The evidence of the rocks showed a history of life "characterized by *rapid* evolutionary events punctuating a history of stasis."[26]

Eldredge and Gould wished to retain the core principle of Darwinian natural selection—the culling of advantageous bodily forms and behaviors by a ruthless environment, and the consequent preservation of their genetic causes, resulting in the spread of the "fit" forms and the extinction of the unfit. But they dramatically altered the pace and mode of change: "The history of evolution is not one of stately unfolding, but a story of homeostatic equilibria, disturbed only 'rarely' (i.e., rather often in the fullness of time) by rapid and episodic events of speciation."[27]

By arguing that the Darwinian paradigm should be retained at the level of genetic change but discarded and replaced at the level of phyletic change, Eldredge and Gould were suggesting considerably more than a bit of tweaking to the Neo-Darwinian Synthesis. They were asserting that the macro level of evolution should be decoupled from the micro level. This formulation had several important implications for the study of the history of life.

First, implicit in their proffered model was the suggestion that most of the large-scale patterns of life's history could not be inferred from the principles at work at the genetic level. Generally, evolutionary change occurred when a new species branched off from a parent species. The great bulk of change did not take place, as the orthodoxy had assumed, within each species. There would have to be new and different techniques invented for studying the causes and consequences of speciation; the work that had been done up to 1972 was mostly irrelevant.

Second, the theory implied that species were real entities—"natural kinds" in the philosophers' parlance—rather than frozen moments in a transience. Evolutionary biologists would no longer be able to model gene flows in populations without reference to the organisms that constituted those populations. Models of speciation would have to recognize that gene flow within species was easy, but that there were boundaries—lines of resistance—at the margins of the species that would have to be overcome if a new species were to come into existence. Once a margin was passed, and a new species created, gene flow could become unrestricted again.

Third, if macroevolution could no longer be inferred from microevolution, then a seemingly discredited bit of theory would have to be revived. Over decades the Synthesists had gradually disconfirmed the notion that groups could be the unit of selection. It had seemed to some ethologists that there were aspects of behavior that could only be explained with reference to an individual sacrificing himself for the good of the group. One member of a flock of birds, for example, might give a warning call upon seeing a hawk in the sky. The call would help to ensure the survival of the other members of the flock, but by calling attention to the one individual that had given the call, it lowered his probability of survival. Would not such behavior have to be explained in terms of the selection of groups? By 1972, the orthodoxy had, with theory and evidence, discredited such theoretical speculations. All behavior had to be explained as benefiting individual organisms, or one or more of their genes.

But if the theory of punctuated equilibria was correct, then the notion of group selection—at least at the species level—would return to relevance within evolutionary biology. The environment had to be seen as culling, not just individuals, but species, all of whose members shared certain characteristics. Why, for example, did the dinosaurs vanish 65 million years ago? Was the disappearance better explained as a failing by each individual of each species, or as a vulnerability shared by all dinosaurs? Thus, punctuated equilibrium raised the possibility of a hierarchical theory of evolution, one that might require different theories for different levels of change.

Together, these differences and implications made the Eldredge and Gould formulation at the least a substantial modification of orthodox Darwinism and at the most a new paradigm of evolution.[28] When the "punctuated equilibrium" essay was published, it provoked a reaction that instantly made disciplinary superstars of Eldredge and Gould. Some paleontologists were converted, and began research projects to illustrate the greater conceptual power of punctuated equilibrium over phyletic gradualism.[29] The greater part of the scientific commentary on the new, alternative paradigm was hostile, however, much of it hysterically so. Some of this criticism continued for a good quarter century. It might be suggested that no scientific theory has been refuted so many times, with such enthusiasm, for so long, as the theory of punctuated equilibrium. The *scientific* criticism, as opposed to its political subtext, was of several types, and warrants summary.

Some of the criticism consisted of picking out individual quotations from Darwin, or from biology textbooks from the 1950s, as part of a campaign to discredit Eldredge and Gould's portrayal of the phyletic gradualism of orthodox Darwinism.[30] As I have mentioned, this sort of attack is misleading and uninteresting.

A more substantive but tractable criticism addressed itself to the operational definition of "punctuation." How long could a period of speciation last before it stopped being an example of a punctuation and started to become an example of phyletic gradualism? At first, the different training and perspectives of paleontologists (who inferred speciation from fossils) and integrative biologists (who worked with living species in real time) led them to assume wildly different definitions of "rapid" and "gradual" evolution. At a

1980 interdisciplinary convention on macroevolution, population geneticists were astonished to learn that spans of time they considered almost an eternity were considered by paleontologists to be a "geological nanosecond."[31]

It took a good while to overcome this sort of confusion, but eventually the authors of the theory settled on operational definitions of a punctuation. Eldredge decided that 5,000 to 50,000 years were a "comfortable yardstick" for a punctuated speciation, while Gould endorsed the modal figure of 40,000 years.[32]

An important line of criticism took off from the assertion that Eldredge and Gould had confused *gradualism* with *constant speedism*. In Richard Dawkins's view, Eldredge and Gould had accused orthodox Darwinists of portraying the evolution of any given lineage as looking, if portrayed graphically, like a ramp. With a constant slope representing phenotypic change, species rose smoothly from one species level to another. According to Dawkins, Eldredge and Gould announced that a much better graph of the pattern of evolution would be a set of stairs. As the steps are level for a period, species experience stasis for a while—and the while, of course, can be a very long time. Then a lineage jumps up to a new level—a new species—after which it is again static for awhile. While Dawkins conceded that the punctuationists' emphasis on phenotypic stasis was a useful contribution to biological discourse, he rejected the idea that it was a significantly different way of conceptualizing change in the history of life. Average out the variations in height between steps, and you get a smooth line—a ramp. "Punctuationists," Dawkins concluded, "are really just as gradualist as Darwin or any other Darwinian; they just insert long periods of stasis between spurts of gradual evolution."[33] Philosopher Daniel Dennett made a very similar criticism (the metaphors of the staircase and ramp are his) by interpreting Eldredge and Gould to have said that the manner in which they challenged gradualism "was not, in the end, by positing some exciting new *non*gradualism, but by saying that evolution, when it occurred, was indeed gradual—but most of the time it was *not even* gradual; it was at a dead stop. . . . Once we get the scale right, they are gradualists themselves."[34]

This sort of criticism is surprisingly misdirected. A line can be drawn through any set of data points, thereby smoothing out the display of information. Because Dawkins and Dennett are speaking schematically to make a point rather than analyzing actual information, they use the simplest possible idea of a line, an average. Dennett uses the metaphor of the ramp to discuss his line, but the principle is the same. When analyzing actual evidence, scholars sometimes use regressions, or logarithms, or even more sophisticated means of drawing lines. But regardless of how schematic or sophisticated the line, it derives its meaning from the variance of the points around the central tendency measured and illustrated by the line. If there is little variance, the line describes a significant relationship. If the variance is wide, there is no evident relationship between the data points, and the line is not helpful in telling us about reality.

The terms *phyletic gradualism* and *punctuated equilibrium* refer to contrasting theories about the rates at which organisms typically create new species. The former predicts that if a line is drawn representing the "average" speciation rate in the history of life, the variance around the line will be small, and the metaphor of the ramp will be appropriate. The latter predicts that the variance around the line will be large, and the metaphor of the stairs will be appropriate.

Measuring speciation rates is fraught with many serious methodological problems, most notably the extremely unrepresentative nature of the fossil record. But given that modern biologists have established that species can stay in stasis for a very long time, and then give birth to new species in remarkably short order, it seems cavalier to characterize the entire record as exhibiting overall gradualism. On one end of the scale, several small crustacean species of the order *Notostraca* seem to have survived, with only trivial differences in their phenotypes, for at least 300 million years.[35] A type of fish living today in the Indian Ocean, the coelacanth, has apparently existed in evolutionary stasis for at least 65 million years, and perhaps for as many as 250 million.[36] The common tree squirrel *Sciurus* has remained in stasis for 35 million.[37] At the other end of the scale, two races of the apple-maggot fly, *Rhagoletis pomonella*, have achieved reproductive isolation in about 150 years, and therefore seem to be in the process of speciation.[38] Lake Nabugabo, formed in Africa, according to geologists, only about 4,000 years ago, has given rise to several dozen species of cichlids, a small fish.[39] The dwarfed woolly mammoth of Wrangell Island appears to have evolved during the Quaternary epoch in 5,000 years, and the dwarfed red deer of the Isle of Jersey in 6,000.[40] A family of *Laupala* Hawaiian crickets seems to have thrown off a new species every 240,000 years.[41] To lightly pass these large variations off as all illustrating gradualism, "once we get the scale right," is not convincing. The admittedly fragmentary evidence suggests that there is a great deal of stasis in life, but that there are undoubtedly moments of relatively brief punctuation.

In a related criticism, orthodox Darwinists noted the emphasis Gould and Eldredge placed upon speciation as occurring during punctuational events, correctly inferred that the new theory thus reintroduced the concept of species selection into the discussion, and declared it off point. Dawkins wrote:

> What I mainly want a theory of evolution to do is explain complex, well-designed mechanisms like hearts, hands, eyes, and echolocation. Nobody, not even the most ardent species selectionist, thinks that species selection can do this. . . .
>
> [Species selection associated with punctuations] . . . could account for the pattern of species existing in the world at any particular time. It follows that it could also account for changing patterns of species as geological ages give way to later ages, that is, for changing patterns in the fossil record. But it is not a significant force in the evolution of the complex machinery of life.[42]

As Gould pointed out, however, Dawkins is here conceding the power of his and Eldredge's explanatory theory, while asserting that it is nevertheless trivial because it does not explain the things Dawkins thinks are worth explaining. Since it is Gould's purpose to explain "the pattern of species existing in the world at any particular time," however, rather than the adaptive complexity that Dawkins believes should be the whole substance of evolutionary theory, Gould is correct in rejecting the charge of triviality as itself trivial.[43]

Another common criticism of the new way of looking at the history of life, however, was neither trivial nor misdirected. It addressed itself to the rather cavalier manner in which Eldredge and Gould had derived evolutionary stasis from lack of evidence of change in the fossil record, and then derived punctuational change from the place where the record of one stasis left off and another began. As a variety of observers within evolutionary biology pointed out, lack of evidence within the fossil record did not constitute evidence of stasis. In Philip Gingerich's words, "Absence of evidence is absence of evidence, not evidence of absence. . . . Punctuated equilibria as an alternative to gradualism is not supported by gaps in the fossil record: gaps are gaps, providing no evidence of transition."[44] There was no way for

Eldredge and Gould to attack the logic of this criticism, or complain that it misunderstood their argument, or maintain that it was inconsistent with observation. They had to hope that their fellow scientists felt that it was a plausible argument that lack of transition fossils in a rock strata could best be interpreted as illustrating a punctuated pattern of species change rather than as simply missing evidence. In the decades since the famous article was published it appears that some biologists are inclined to equate gaps in the evidence with evidence of rapid change, while most are not.

The main line of scientific attack on the theory of punctuated equilibrium, however, one that activated an enormous amount of research effort within paleontology, was empirical. Gould and Eldredge, separately and together, Steven Stanley, Elizabeth Vrba, and others produced new research and reanalyzed old research to illustrate their contention that the story in the rocks validated a notion of stasis and punctuation.[45] Meanwhile, over decades, and continuing after Gould's death in 2002, other paleontologists produced evidence that they said showed a predominant pattern of gradualistic evolution, thus refuting the theory of punctuated equilibrium.[46] Some researchers found evidence of both gradualist and punctuationist patterns in the imperfect record of the rocks.[47]

Needless to say, this war of empirical evidence was accompanied by a great deal of methodological wrangling. How should we define the concept of "species" so that we know we are arguing about the same conceptual entities? (There are nine separate definitions of the concept in common use among integrative biologists.)[48] How should "punctuation," "gradual," and "rapid" be defined and measured? How can a sudden speciation, in the record of the rocks, be differentiated from the record of an in-migration? How can species that vary in behavior but are identical in morphology be differentiated? How can families as unequivalent as mammals and mollusks be arrayed along a similar time line so that their speciation and extinction rates can be compared? Does it make sense to apply the same conceptual framework to plants and animals, when we know that they reproduce through very different genetic mechanisms? How important is it to differentiate the type of speciation that takes place under "normal" earth conditions of competition among individuals, and the type of speciation that occurs when competition temporarily disappears, such as occurs when survivors face an underpopulated world after a mass extinction, or an evolutionary innovation projects a lucky species into a new type of environment (when animals moved from water to land, for example)? And, always and everywhere among paleontologists, how do we correct for the extreme imperfection of the fossil record?

Reviewing more than three decades of controversy, the best this nonspecialist can say is that the outcome is still in doubt, but the belief that the punctuational model, with its attendant concepts of species selection (or, as Eldredge and Gould preferred to term it, *species sorting*) and hierarchical evolution, provides a better alternative to the understanding of the history of life than orthodox Darwinism is endorsed by a clear minority of paleontologists and integrative biologists. Gould may have convinced many readers outside of science that his theory best represented the reality of life's history, but he did not convert most of the members of his own and related scientific professions.

The scientific arguments are only part of the story, however, and not necessarily the most interesting part. Almost from the first moment, the idea of punctuated equilibrium was seen, by those inside and outside of science, to have political implications.

To Christian creationists, Gould, with his lucid prose style and deliberate targeting of the nonscientific community, and his critical take on orthodox Darwinism, was, well, a godsend. Esoteric and jargonized as most professional publications were in evolutionary biology, they did not lend themselves to understanding by uneducated outsiders. But much of Gould's explanation of his and Eldredge's theory could be understood by any intelligent person. And, once they were able to understand the theory, the members of the anti-science fraternity were perfectly capable of concluding that *absence of evidence* among fossils might be construed as *evidence of miraculous creation.* Creationists pounced. "Harvard scientists agree: Evolution is a hoax!!!" read one broadsheet that crossed Gould's desk.[49] "The experts know transitional forms are, in general, a myth. And sometimes, as in the case of Stephen Jay Gould, they even admit it. . . . I see PE as the one evolutionary theory that tries to remain true to the fossil record," wrote one creationist after Gould's death.[50] Gould's line about the absence of transitional forms being the trade secret of paleontology is quoted endlessly by creationists.[51]

Gould, of course, attempted to squash the notion that any of his ideas in any way provided support for creationist theories, protesting the use of his own words for "creationist pseudoscientific practice."[52] "It is infuriating to be quoted again and again by creationists—whether through design or stupidity, I do not know—as admitting that the fossil record includes no transitional forms," he wrote in 1983. "Transitional forms are generally lacking at the species level, but they are abundant between larger groups."[53] But despite his protests, repeated many times over his career, creationists went right on using his early words against the science he loved.

Gould cannot be blamed for supplying creationists with a weapon they were happy to use to undermine Gould's own profession. Any theory is apt to be misunderstood and misapplied by later readers, and its author is not responsible for the results of other people's corruption of his ideas. Still, Gould did make it easier to misunderstand by succumbing to the temptation to overuse his rhetorical skills. In both his professional publications and popular writing, his early rhetoric about visible stasis and invisible speciation had emphasized the absence of transitional forms. The qualifier about the difference between levels did not begin appearing in his work until he realized what the creationists were doing with his theory. Reading the early writings, it would be easy to get the impression that all transitions, at all levels, are absent from the fossil record. By the time Gould realized that his rhetoric, fashioned for one purpose, was being hijacked for another, it was too late to edit his words.

The creationists were an annoying distraction, but Gould had more important political arguments to push. Running through the original "punctuated equilibrium" model had been a philosophic concern with the directing, and sometimes confining, functions of dominant paradigms. In the original 1972 paper Eldredge and Gould wrote that "all observation is colored by theory and expectations," discussing a few philosophers of science to buttress the claim.[54] The imaginations of evolutionary biologists, they asserted, had been so dominated by the gradualism of Darwin and the Synthesis that they had failed to properly interpret the evidence in front of them. Once their minds were freed from the misleading paradigm, they would be able to see the record of the rocks in a new way.

The original paper had contained only a hint that the new scientific paradigm might also be political. But in 1977 Gould wrote, with Eldredge as second author, an assessment of the new paleontological evidence relevant to patterns of evolution. He returned to the issue of the channeling effects of scientific paradigms, and this time the connection between scientific and political world-views was explicit: "The general preference that so many of us hold for gradualism is a metaphysical stance embedded in the modern history of Western cultures." He then quoted with approval William Irvine, an "astute biographer" of Darwin, who had opined, "The economic conception [of laissez-faire liberalism] . . . can all be paralleled in the *Origin of Species*."[55]

In many other publications, Gould repeated the point that Darwin's theory was "the economy of Adam Smith transferred to nature."[56] On the one hand, Darwin, like Smith, argued that the struggle of individuals made for the advancement of the whole, and Darwin, like Smith, emphasized that this was a long, slow process of improvement. Darwinism, and therefore the Synthesis, thus prepared the minds of scientists—and therefore the minds of the scientifically aware public—to value gradual amelioration and oppose abrupt transformations in society. On the other hand, a theory of punctuated equilibrium was consistent with conceptions of historical change offered by Hegel, Marx, and Engels.[57] Not then, or ever, did Gould argue that embracing punctuationism in scientific discourse would necessarily lead to a dialectical mode of political thinking. That would have been nonsensical, because his coauthor, and the originator of the concept, was not a Marxist and did not come from a left-wing background. (In fact, Eldredge's family was Republican.)[58] But Gould managed to leave the impression that the adoption of the punctuationist paradigm in science might help dissolve the mental shackles that kept scientists, and citizens, from appreciating the value of a rapid transformation in society.

Moreover, in another article in 1980, solo authored, Gould wrote, "I believe that, in ways deeper than we realize," the preference of scientists for slow, low-level changes in organic lineages as the basis for high-level changes (that is, for micro as opposed to macro thinking) "represents a cultural tie to the controlling Western themes of progress and ranking by intrinsic merit."[59] The same year, in one of his popular essays, he wrote that "gradualism is a culturally conditioned prejudice, not a fact of nature."[60] In other words, phyletic gradualism is one facet of a more general tendency in capitalist culture to see inequalities as embedded in nature, rather than as artificially created by society.

Therefore, in Gould's view, normal evolutionary biology, while it did not create a political ideology, facilitated the ideology in place. In contrast, a punctuational view of the history of life, while it would not mandate a change in mass political ideology, might prepare the ground for such a change. Viewed from one perspective, science and politics are separate entities. Viewed from another, they are mutually reinforcing aspects of an ideological whole.

The Theory with a Life of Its Own

Almost as soon as the punctuated equilibrium theory appeared in print, various thinkers realized that in its general form it provided a model for stability and change in any complex system. As an ideal pattern existing in abstraction rather than a specific theory of natural selection, it could be adapted to a seemingly limitless sample of natural and human phenomena. There is always resistance to change, and the resistance prevails for a time, resulting in stasis. After awhile, the forces tending toward change build to an irresistible level, at which point change is rapid and thorough. Then stasis prevails again until next time. Thus, the general punctuational model has been applied to human learning, organizational dynamics, technological development, fractal geometry, chaos theory, nonlinear dynamics, complexity theory, economics, and other realms.[61]

Eldredge has embraced the applications of punctuated equilibrium to human history, suggesting that "social entities may themselves be susceptible to a form of selection directly analogous to large-scale natural selection."[62] Gould, however, has been more cautious in endorsing the promiscuous application of their theory. While pronouncing himself pleased to have been the inspiration for so much fruitful thought,[63] he also feels the obligation to point out the potential dangers in the enterprise.

First, the biological theory of punctuated history is strictly Darwinian in that it conceptualizes change in life over time as entirely a process of the conservation of randomly generated mutations. Organisms do not inherit acquired modifications. As Gould points out often, however, culture is Lamarckian: the inheritance of acquired characteristics is one of its defining features.[64] Thus, the addition of consciousness and meaning to human history complicates the process of analyzing change.

Second, and more important, in wrenching a specific theory out of its specific context, other intellectuals have loosed it from its empirical anchor. Despite his rather imprecise hopes that the theory might help to change political ideologies, at the level of science Gould is a committed empiricist, and he is wary of grand theoretical projects not safely secured to operational definitions. When punctuated equilibrium is removed from the empirical base where Eldredge and Gould situated it, it loses its character as a theory and becomes only a metaphor. Gould warns that "these 'brave' statements about conceptual homology across disparate scales and immediate causalities must remain empty and meaningless without operational criteria for distinguishing . . . meaningful similarities of genesis (homology) from misleading superficiality of appearance (analogy). . . . I am more interested in exploring ways in which the theory might supply truly causal insights . . . rather than broader metaphors that can surely nudge the mind into productive channels, but that make no explicit claim for causal continuity or unification."[65]

This caution has direct application to any consideration of Gould as a socially relevant theorist. For the idea of punctuated equilibria has had an impact in the humanities and social sciences, also. But the way it has been assimilated suggests that Gould's distinction between theories and metaphors has not always been appreciated.

A good example is a 1996 effort at big-picture prognosticating by economist Lester Thurow, *The Future of Capitalism.* Pointing out that the world seems to be in a period of economic and political turmoil, Thurow promises to make the future understandable by borrowing "two concepts from the physical sciences—plate tectonics from geology and punctuated equilibrium from biology."[66] The difference between a concept, which might have an operational definition, and a metaphor, which is wholly verbal, is thus fuzzed at the beginning. The initial vagueness is enhanced by Thurow's lack of knowledge of the scientific origins of his metaphors. He seems to have read nothing of the geological literature on plate tectonics. His source for the details of the idea of punctuated equilibrium is an article by another economist.[67] Not surprisingly, he garbles the idea of a punctuation, summarizing it thus: "The environment suddenly changes and what has been the dominant species rapidly dies out to be replaced by some other species. . . . The best-known example, of course, is the dinosaurs."[68]

Having thus appropriated a metaphor with a faux-scientific pedigree, Thurow can then inform his readers, "Periods of punctuated equilibrium are equally visible in human history," and freely indulge his imagination.[69] It turns out that "today the world is in a period of punctuated equilibrium—which is being caused by the simultaneous movements of five economic plates."[70] The rest of the book, while it contains some perceptive discussions of contemporary economic problems, has nothing to do with evolutionary biology. Apparently, the point of using the idea of punctuated equilibria was to add some rhetorical pizzazz to an otherwise fairly standard liberal-left analysis of American economic problems. The exercise was harmless enough, but it did risk giving readers the impression that there was more natural-science heft behind a set of ideas than was actually there.

The tendency to want to recruit the presumed intellectual prestige of biology to lend weight to a set of social science concepts is also on display in the discipline of political science. Here, the greatest influence of the theory has been in its application to the "realignment" or "critical elections" conceptual framework. Arising out of a famous article by V. O. Key in the 1950s,[71] it was worked into a coherent, wide-ranging, and extremely influential theory of American elections by Walter Dean Burnham, James Sundquist, and others.[72] It envisions American electoral history as being divisible into eras, each of approximately a generation in length. During stable periods, the coalitional bases of the two major parties remain relatively impervious to various stresses arising from developmental forces in the economy and society. After roughly thirty years, however, the stresses have built to a breaking point, and some crisis (war or civil war, depression, etc.) arises that causes social discontent to boil over and dissolve the psychic bonds that attach citizens to parties. In one or a series of critical elections (1800, 1828, 1856–1860, 1896, 1928–1936, and perhaps 1968–1980), new issues and leaders arise, party coalitions are reshuffled, a new electoral party majority is created that, once in power, enacts new policies, and the new political arrangement endures for about another generation.

Although the critical elections framework has been enormously influential in organizing vast quantities of research by political scientists and historians, it has attracted persistent skepticism. Among other failings, critics fault its alleged lack of clear operational definitions for both stasis and realignment, its supposed inability to generate unambiguous research, its lack of a consistent explanation for the period since 1964, in which party decomposition rather than party realignment seems evident, and its failure to integrate microtheories of voting behavior with its own pretensions of being a macroexplanation of American politics.[73] In response, Walter Dean Burnham, the foremost expositor of the theory, has, in effect, defended it by arguing that his formulation is only a subcategory of the larger pattern of historical change posited by natural science. Burnham no longer speaks of critical realignments, but of punctuated equilibrium, with explicit reference to Eldredge and Gould.[74]

My point here is not to attack or defend the notion of critical elections as a useful insight into American politics. My purpose is to point out that the application of the concept of punctuated equilibria to electoral history is a perfect example of the confusion of empirical homology with nonempirical analogy against which Gould warned. In biology, the theory is based upon measurable trends in the fossil record and backed by an explicit model of historical process. In political science, it rests uneasily upon a debatable record of voting and policy enactment, an even more debatable series of assumptions about the relationships among socio-economic trends, voting, and policy, and no model of historical process. It is, in short, a metaphor, not a theory. It may be a good metaphor, but it has no causal explanatory power of the type Gould insisted upon for its use.

The same can be said about other applications of punctuated equilibrium in political science. For example, Baumgartner and Jones base their study of policymaking in American politics explicitly on a punctuationist model. Citing the original Eldredge and Gould article and several others of their individually authored formulations, Baumgartner and Jones assert, "Punctuated equilibrium, rather than stability and immobilism, characterizes the American political system."[75] As with discussions of critical elections, however, their formulation has no grounding in operational definitions of stasis and sudden change, and is therefore a metaphor rather than a causal theory. The authors actually state, "We have adopted the terminology of punctuated equilibrium because it evokes the image of stability interrupted by major alterations in the system."[76] Clearly, any framework that relies upon evocative images rather than operational definitions of core concepts is a metaphor at heart.[77]

Political science thus looks to Eldredge and Gould for inspiration, not for a model. Their scientific theory has become a handy window-dressing for social scientists looking to borrow the cachet of natural science while remaining metaphorical in fact.

The Contingency of History

Gould's writings on history were not confined to the subject of punctuated equilibrium. Another theme that he explored often is that history is a "contingent" science, by which he meant that although life evolves according to general laws, the individual working out of myriad causes renders prediction of specific outcomes impossible. He defined the term *contingency* as "the tendency of complex systems with substantial stochastic components, and intricate nonlinear interactions among components, to be unpredictable in principle from full knowledge of antecedent conditions, but fully explainable after time's actual unfolding."[78] Thus, "the unique contingencies of history, not the laws of physics, set many properties of complex biological systems," and "contingency represents the historian's mode of knowability."[79] Nothing can be predicted, but everything can be explained in retrospect.

Gould's most extensive elaboration of this theme is found in his 1989 book *Wonderful Life*, a description of and a philosophical argument based on the Burgess Shale, a deposit of 500-million-year-old soft-bodied animal fossils discovered in British Columbia in 1910. These animals lived at the beginning of what paleontologists call the "Cambrian Explosion," a sudden (as these things are reckoned in geology) efflorescence of multicelled creatures following on about 3.3 billion years when single-celled organisms were the only life on earth. According to Gould, many of these animals are so peculiar in body type that they could not be fit into any of the twenty or so phyla (major categories) that organize the kingdom of Animalia now. "The Burgess Shale . . . contains the remains of some fifteen to twenty organisms so different one from the other, and so unlike anything now living, that each ought to rank as a separate phylum."[80]

This extraordinary diversity of form at the beginning of multicellular life was significant, Gould suggested, because it contradicted the expectations derived from orthodox Darwinism. As he portrayed orthodoxy, its story was that life evolved from simple to complex, and restricted to diverse, with more sophisticated creatures always evolving until the most sophisticated—*Homo sapiens*—arrived as a sort of culmination. In this story, we, the most evolved, were inevitable. The moment that a single molecule began to replicate in the primal ooze, human beings, or something very much like them, were bound to evolve unless some cosmic catastrophe exterminated all life before it could advance. Gould called this the "tendency theory" of human evolution: "*Homo sapiens* becomes the anticipated result of an evolutionary tendency."[81]

But the Burgess Shale teaches us that "the story of the last 500 million years has featured restriction followed by proliferation within a few stereotyped designs, not general expansion of range and increase in complexity as our favored iconography, the cone of increasing diversity, implies."[82] The conclusion must be that evolution follows no inevitable path. Therefore, humans are the result of a concatenation of accidents, and not in the least inevitable. "We are an improbable and fragile entity, fortunately successful after precarious beginnings as a small population in Africa, not the predictable end result of a global tendency."[83]

If this is so, then the larger truth is that evolution has proceeded largely according to luck. The Darwinian concept of the "fit" has always meant that some species, or individuals within species, were better adapted to the environment. In that sense, "survival of the fittest," although it was not a judgment of moral worth, was definitely a judgment of functional superiority. But, Gould argued, "if we face the Burgess fauna honestly, we must admit that we have no evidence whatsoever—not a shred—that losers in the great decimation were systematically inferior in adaptive design to those that survived."[84] Apparently, animals survived or perished according to random fortune. The metaphor of the lottery recurs in *Wonderful Life*.[85] Thus, the history of life is not the story of the survival of the fittest, but of "massive historical contingency."[86]

If contingency dominates evolution, "Wind back the tape of life to the early days of the Burgess Shale; let it play again from an identical starting point, and the chance becomes vanishingly small that anything like human intelligence would grace the replay."[87] "Each replay of the tape would yield a different set of survivors and a radically different history."[88] The species *Homo sapiens* is not the crown of creation; it has, improbably, won the evolutionary lottery.

It is not at all clear that Gould correctly portrayed the views of orthodox evolutionary biologists in this book. Many have since disavowed the views attributed to them in *Wonderful Life*. Nevertheless, as I will explain shortly, some natural scientists do hold this view. Whether the opinions Gould tried to discredit were actually representative of most in his profession, or constituted a straw man he constructed for convenience, his argument is worth examining.

Two political implications derive from his line of argument. I have discussed the first in chapter 2: humans, being a cosmic accident rather than the most evolved species on the planet, have no cause to feel superior to any other species. Indeed, if we measure "superiority" in biological terms by longevity, ubiquity, usefulness to other forms of life, or total biomass, bacteria are the most important kind of life on earth.[89] We are not living in the age of humans, or of mammals, or of animals, or of multicellular life, as legions of popularizers have proclaimed for centuries. Instead, "We live now in the 'Age of Bacteria.' Our planet has always been in the 'Age of Bacteria.'"[90] To repeat, this argument is part of a political strategy to convince people that they should respect other species more by convincing them that they are not better than other species.

The second political implication derives from the connection between contingency and equality. The conventional, conservative Social Darwinist line of argument goes something like this: Just as people deserve to exercise dominion over the other species because they are the most evolved, just so the dominant classes deserve to exercise dominion over the others because they are the fit in social evolution, and the white race deserves to exercise dominion over others because it is the fittest within the grouping of *Homo sapiens*. The argument rests on the premise that those in a dominant position have acquired their status through some sort of superior performance, that is, through merit. Gould seeks to puncture this ideological balloon by deflating the premise, by substituting the idea of chance for the idea of merit at every level of dominance. His rhetorical strategy rests on the use of the metaphor of the lottery, for no one believes that any notion of deservingness attaches to the fact of having won a raffle.

The general, large conclusion—that no one *deserves* their lot in life, either among the elite or among the mass, and that therefore redistribution is justified—is left largely implicit. But in his various essays, Gould leaves little doubt that his goal is to encourage as much equality as possible. His left-wing Social Darwinism—inferring the social imperative of equality from the facts of nature—is a chief conclusion of many of these essays:

> Human equality is a contingent fact of history.[91]
>
> Elitism is repulsive when based upon external and artificial limitations like race, gender, or social class. Repulsive and utterly false—for that spark of genius is randomly distributed across all the cruel barriers of our social prejudice.[92]
>
> Most of hominid history has featured a bush, sometimes quite substantial, of existing species. The current status of humanity as a single species, maximally spread over an entire planet, is distinctly odd. But if modern times are out of joint, why not make the most of it? . . . We could do it; we really could. Why not try sistership; why not brotherhood?[93]

> Life runs more like a caucus race than along a linear course with inevitable victory to the brave, strong, and smart. If we truly embraced this metaphor in conceptual terms, we might even be able to adopt a better position for considering the moral consequences of human actions, as suggested by Lewis Carroll's wise dodo: no judgments of superiority or inferiority among participants; no winners or losers; and cooperation with ends attained and prizes for all.[94]

As with the theory of punctuated equilibrium, Gould's interpretation of life as being an exercise in contingency drew both scientific and political reactions, and it started a disciplinary argument that continues today. At the level of science, the idea of the lottery-like nature of survival in evolution drew a surprising amount of support from paleontologists. Ernst Mayr, who did not hesitate to disagree with Gould on other issues, concurred on this one: "The more I study evolution the more I am impressed by the uniqueness, by the unpredictability, and by the unrepeatability of evolutionary events."[95] George Gaylord Simpson had made essentially the same claim.[96]

But there were many who disagreed. Before Gould was born, Julian Huxley had presented the case for a progressive essence to evolution.[97] After *Wonderful Life* appeared, Richard Dawkins expressed a view contrary to Huxley's, but also contrary to Gould's. Gould, Dawkins suggested, had misunderstood the Burgess Shale:

> The true index of how unalike two animals are is how unalike they actually are. Gould prefers to ask whether they are members of the known phyla. But known phyla are modern constructions. . . . The five-eyed, nozzle-toting *Opabinia* cannot be assimilated to any textbook phylum. But, since textbooks are written with modern animals in mind, this does not mean that *Opabinia* was, in fact, as different from its contemporaries as the status "separate phylum" would suggest. . . . The view that he is attacking—that evolution marches inexorably towards a pinnacle such as man—has not been believed in years.[98]

Similar criticisms have been written by Daniel Dennett, Timothy Shanahan, Jeffrey Levinton, and Edward O. Wilson.[99] All of these targeted their attacks directly at Gould's discussion of the Burgess Shale and what it meant for the pattern of life's development, ignoring the political implication of his argument.

Paleontologist Simon Conway Morris, however, took on both Gould's science and his politics. Morris had actually worked extensively with the fossils of the Burgess Shale. In 1998 appeared *The Crucible of Creation*, his refutation of Gould's science. Morris argued, first, that Dawkins was right: "The strangeness of the problematic Cambrian animals is really an artifact, a construction of our imagination."[100] Second, Gould was wrong about the arc of life being first luxuriant, then constricted. The standard Darwinian image of ever-increasing diversity through geological time is correct.[101] As with all other disputes about life's history, the fragmentary nature of the evidence about the Burgess Shale prevented even the word of an expert such as Morris from being definitive. But inasmuch as the consensus of evolutionary scientists counted as evidence itself, it appeared that Morris spoke for the majority, and Gould's views were once more representative of only a few.

Morris made a few short references to his additional disagreement with Gould's ideological preferences, but his philosophical riposte required another book. In *Life's Solution* Morris argued that the phenomenon of "convergence" proved that, while contingency might govern the small things in the short run, in the long run the direction of evolution was entirely predictable. Morris elaborated on "the recurrent tendency of biological organization to arrive at the same 'solution' to a particular need" throughout the animal world.[102] Silk-weaving has been "invented" by spiders, fungus gnats, midges, caddis-flies, weaver ants, and, of course, silkworms.[103] Eyes have evolved independently in squids, snails, polychaetes, jellyfish, arthropods, dinoflagellates, vertebrates, and other groups.[104] Ears with tympanic membranes have evolved independently several times among the vertebrates.[105] Humans are clearly the most intelligent organisms, but something approaching conceptual intelligence has evolved in dolphins, monkeys, parrots, crows, and even perhaps in army ants.[106] Thus, "If we had not arrived at sentience and called ourselves human, then probably sooner rather than later some other group would have done so.[107]

Therefore, Gould's image of an evolutionary process governed by chaos and accident is completely wrong. "The ubiquity of convergence will lead inevitably to the emergence of recurrent biological properties that define the fabric of the biosphere. Rerun the tape of life as often as you like, and the end result will be much the same."[108]

But as with Gould, the point of the examination of the evidence of life is not a sterile understanding of science, but an appreciation of its implications. Whereas Gould's affirmation of the ruling concept of chance led him to conclusions about merit and equality, Morris had more transcendent purposes. Morris was not one to claim that science and morality occupy different spheres. Quite the contrary: "Societies that ignore what we discover do so at their peril, but if they imagine that on occasion the discoveries of evolution are neutral in their implications, again societies delude themselves."[109] And what we have discovered, Morris said, was that the "salient facts of evolution are congruent with a Creation. . . . None of it presupposes, let alone proves, the existence of God, but all is congruent."[110]

It is my own view that Gould, in his rhetorical strategy of implication, and Morris, in his strategy of explication, are both right. Ideological arguments and conclusions are not adventitious adjuncts to the process of biological science; they are part of its fabric. Every interpretation of life's history is linked to moral, philosophical, and political implications. It is possible to ignore them, of course. When Dawkins, Wilson, and others criticized Gould's book on the Burgess Shale, they confined themselves to attacking his science. But by omitting the political point of view that accompanied the scientific point of view, they inevitably gave a misleading account of Gould's argument. Morris, who frankly admitted his ideological hostility as part of his scientific critique, thereby gave a far more satisfying account of the controversy.

The lesson to be learned from Gould's forays into the history of life is thus not that he was correct or incorrect. It is that the right interpretation of facts is only part of the process of investigating the past.

notes

1. J. Daryle Charles, review of Richard J. Evans, *In Defense of History, Academic Questions* (Winter 2000): pp. 92–94.
2. Leonard Krieger, *Time's Reasons: Philosophies of History Old and New* (Chicago: University of Chicago Press, 1989), p. 11.
3. Christopher Lloyd, *The Structures of History* (Cambridge, MA: Blackwell, 1993), pp. 17, 47, 164.
4. Krieger, *Time's Reasons*, p. 2.
5. Stephen Jay Gould, *The Structure of Evolutionary Theory* (Cambridge, MA: Harvard University Press, 2002), p. 1183.

6. Stephen Jay Gould, *Wonderful Life: The Burgess Shale and the Nature of History* (New York: W. W. Norton, 1989), p. 323.

7. Charles Darwin, *On The Origin of Species*, 1st ed. (New York: Barnes and Noble, 1962 [1859]); Charles Darwin, *On The Origin of Species*, 6th ed. (New York: Macmillan, 1962 [1872]); Ernst Mayr and William B. Provine, eds., *The Evolutionary Synthesis: Perspectives on the Unification of Biology* (Cambridge, MA: Harvard University Press, 1998); Peter B. Bowler, *Evolution: The History of an Idea* (Berkeley: University of California Press, 2003), pp. 155–76.

8. E. E. Schattschneider, *The Semi-Sovereign People* (New York: Holt, Rinehart and Winston, 1960), p. 2.

9. I discuss this point in more detail in *The Paradox of Democratic Capitalism: Politics and Economics in American Thought* (Baltimore: Johns Hopkins University Press, 2006), pp. 291–95.

10. Gould, *Wonderful Life*, p. 48.

11. Darwin, *Origin*, first ed., pp. 52, 77.

12. Richard Dawkins, *The Blind Watchmaker: Why the Evidence of Evolution Reveals a Universe Without Design* (New York: W. W. Norton, 1996), p. 264.

13. Ibid., 151.

14. Pat Shipman, *Taking Wing: Archaeopteryx and the Evolution of Bird Flight* (London: Phoenix, 1998), pp. 23–35; Paul Chambers, *Bones of Contention: The Archaeopteryx Scandals* (London: John Murray, 2002), pp. 110, 148, 183–85.

15. R. A. Fisher, *The Genetical Theory of Natural Selection* (Oxford: Oxford University Press, 1930), pp. 67, 78; Theodosius Dobzhansky, *Genetics and the Origin of Species* (New York: Columbia University Press, 1982), pp. 192, 310.

16. George Gaylord Simpson, *Tempo and Mode in Evolution* (New York: Columbia University Press, 1944), pp. 39, 99–114; Ernst Mayr, "Change of Genetic Environment and Evolution," in J. Huxley, A. C. Hardy, and E. B. Ford, eds., *Evolution as a Process* (London: Allen and Unwin, 1954), pp. 157–80, reprinted in Ernst Mayr, *Evolution and the Diversity of Life* (Cambridge, MA: Harvard University Press, 1976), pp. 188–210.

17. George Gaylord Simpson, *The Major Features of Evolution* (New York: Columbia University Press, 1953); Niles Eldredge, *Reinventing Darwin: The Great Evolutionary Debate* (London: Phoenix, 1995), pp. 24–29; Gould, *Structure*, pp. 528–31.

18. Ernst Mayr, introduction to "Change of Environment and Speciation," in Mayr, *Diversity of Life*, p. 188.

19. Niles Eldredge, *Time Frames: The Evolution of Punctuated Equilibria* (Princeton, NJ: Princeton University Press, 1985), p. 75.

20. Niles Eldredge, "The Allopatric Model and Phylogeny in Paleozoic Invertebrates," *Evolution* 25, no. 1 (March 1971): 156–67.

21. Niles Eldredge and Stephen Jay Gould, "Punctuated Equilibria: An Alternative to Phyletic Gradualism," in Thomas J. M. Shopf, ed., *Models in Paleobiology* (San Francisco: Freeman, Cooper, 1972), pp. 82–115; Stephen Jay Gould and Niles Eldredge, "Punctuated Equilibria: The Tempo and Mode of Evolution Reconsidered," *Paleobiology* 3 (1977): 115–51; Stephen Jay Gould and Niles Eldredge, "Punctuated Equilibrium Comes of Age," *Nature* 366 (November 1993): 223–27.

22. Eldredge and Gould, "Punctuated Equilibria," p. 112.

23. Ibid., p. 89.

24. Ibid., p. 122.

25. Stephen Jay Gould, *Ever Since Darwin: Reflections in Natural History* (New York: W. W. Norton, 1977), p. 181.

26. Eldredge and Gould, "Punctuated Equilibrium," p. 108.

27. Ibid., p. 84.

28. Michael Ruse, "Is the Theory of Punctuated Equilibria a New Paradigm?" in Albert Somit and Steven A Peterson, eds., *The Dynamics of Evolution: The Punctuated Equilibrium Debate in the Natural and Social Sciences* (Ithaca, NY: Cornell University Press, 1992), pp. 139–67.

29. Steven Stanley, *Macroevolution: Pattern and Processes* (San Francisco: W. H. Freeman, 1979).

30. Antoni Hoffman, *Arguments on Evolution: A Paleontologist's Perspective* (New York: Oxford University Press, 1989), pp. 102–103; Daniel Dennett, *Darwin's Dangerous Idea: Evolution and the Meanings of Life* (New York: Simon and Schuster, 1995), p. 290.

31. Jeffrey S. Levinton, *Genetics, Paleontology, and Macroevolution* (Cambridge: Cambridge University Press, 2001), p. 348.

32. Eldredge, *Time Frames*, p. 121; Gould, *Structure*, pp. 768, 782.

33. Dawkins, *Blind Watchmaker*, p. 248.

34. Dennett, *Darwin's Dangerous Idea*, p. 290.

35. Jerry A. Coyne and H. Allen Orr, *Speciation* (Sunderland, MA: Sinauer Associates, 2004), p. 425.

36. Dawkins, *Blind Watchmaker*, p. 246.

37. Robert L. Carroll, *Patterns and Processes of Vertebrate Evolution* (Cambridge: Cambridge University, 1997), p. 168.

38. Coyne and Orr, *Speciation*, pp. 159–62, 426.

39. Ibid., p. 419.

40. Gould, *Structure*, p. 852.

41. Tamra C. Mendelson and Kerry L. Shaw, "Rapid Speciation in an Arthropod," *Nature* 433 (January 2005): 375–76.

42. Dawkins, *Blind Watchmaker*, pp. 265, 268–69.

43. Gould, *Structure*, pp. 1020–21.

44. Philip D. Gingerich, "Punctuated Equilibria—Where is the Evidence?" *Systematic Zoology* 33, no. 3 (September 1984): 338.

45. Gould and Eldredge, "Tempo and Mode;" Stanley, *Macroevolution*; Steven Stanley, "Macroevolution and the Fossil Record," *Evolution* 36, no. 3 (1982): 460–73; Elisabeth S. Vrba and Niles Eldredge, "Individuals, Hierarchies and Processes: Towards a More Complete Evolutionary Theory," *Paleobiology* 10, no, 2 (1984): 146–71; Mark A. S. McMenamin, "The Origins and Radiation of the Early Metazoa," in K. C. Allen and D. E. G. Briggs, eds., *Evolution and the Fossil Record* (Washington, DC: Smithsonian Institution, 1990), 73–98; Christopher R. C. Paul, "Patterns of Evolution and Extinction in Invertebrates," in ibid., 99–121; Gould and Eldredge, "Comes of Age;" Niles Eldredge, *Reinventing Darwin: The Great Evolutionary Debate* (London: Phoenix, 1995); Gould, *Structure*, pp. 745–1022.

46. Hoffman, *Arguments on Evolution*, pp. 106–108; John Alroy, "Constant Extinction, Constrained Diversification, and Uncoordinated Stasis in North American Mammals," *Palaeo* 127 (1995): 285–311; Carroll, *Patterns and Processes*, pp. 89, 94, 100, 106; Levinton, *Genetics, Paleontology, and Macroevolution*, pp. 27–29, 143–46, 311–31, 342.

47. Philip D. Gingerich, "Species in the Fossil Record: Concepts, Trends, and Transitions," *Paleobiology* 11, no. 1 (1985): 27–41.

48. Coyne and Orr, *Speciation*, p. 447.

49. Gould, *Structure*, p. 987.

50. Kenneth Poppe, *Reclaiming Science from Darwinism: A Clear Understanding of Creation, Evolution, and Intelligent Design* (Eugene, OR: Harvest House, 2006), pp. 259, 270.

51. For example, Poppe, *Reclaiming Science*, p. 188; Duane T. Gish, *Evolution? The Fossils Say No!* (San Diego: Creation-Life Publishers, 1979), 3rd. ed., p. 176; Michael Denton, *Evolution: A Theory in Crisis* (Chevy Chase, MD: Adler and Adler, 1986), p. 194; Philip E. Johnson, *Darwin On Trial* (Downers Grove, IL: InterVarsity Press, 1993), p. 59; Alvin Plantinga. "When Faith and Reason Clash: Evolution and the Bible," in Robert T. Pennock, ed., *Intelligent Design, Creationism, and Its Critics: Philosophical, Theological, and Scientific Perspectives* (Cambridge, MA: MIT, 2001), p. 133; Ann Coulter, *Godless: The Church of Liberalism* (New York: Three Rivers Press, 2007), p. 219.

52. Gould, *Structure*, p. 989.

53. Stephen Jay Gould, *Hen's Teeth and Horse's Toes: Further Reflections in Natural History* (New York: W. W. Norton, 1983), p. 260.

54. Eldredge and Gould, "Punctuated Equilibria," pp. 85, 90, 91.

55. Gould and Eldredge, "Tempo and Mode," p. 145.

56. Gould, *Ever Since Darwin*, p. 12.

57. Gould and Eldredge, "Tempo and Mode," pp. 145–46.

58. Niles Eldredge in conversation with author, June 21, 2006, New York City.

59. Stephen Jay Gould, "Is a New and General Theory of Evolution Emerging?" *Paleobiology* 6, no. 1 (1980): 126.

60. Stephen Jay Gould, *The Panda's Thumb: More Reflections on Natural History* (New York: W. W. Norton, 1980), p. 195.

61. Gould, *Structure*, pp. 80, 922, 956.

62. Niles Eldredge, "Punctuated Equilibria, Rates of Change, and Large-Scale Entities in Evolutionary Systems," in Albert Somit and Steven A. Peterson, eds., *The Dynamics of Evolution: The Punctuated Equilibrium Debate in the Natural and Social Sciences* (Ithaca, NY: Cornell University Press, 1992), p.118.

63. Gould, *Structure*, p. 923.

64. Gould, *Full House*, p. 222.

65. Gould, *Structure*, pp. 929–30, 952.

66. Lester C. Thurow, *The Future of Capitalism: How Today's Economic Forces Shape Tomorrow's World* (New York: Penguin, 1996), p. 6.

67. Ibid., p. 7; the article Thurow cites is John M. Gowdy, "New Controversies in Evolutionary Biology: Lessons for Economics," *Methodus* (June 1991): p. 86.

68. Thurow, *Future of Capitalism*, p. 7.

69. Ibid.

70. Ibid., p. 8.

71. V. O. Key Jr. "A Theory of Critical Elections," *Journal of Politics* 17 (1955): 3–18.

72. Paul Allen Beck, "A Socialization Theory of Partisan Realignment," in Richard Niemi and associates, *The Politics of Future Citizens* (San Francisco: Jossey-Bass, 1974), pp. 199–219; David W. Brady, *Critical Elections and Congressional Policy Making* (Stanford, CA: Stanford University, 1988); Walter Dean Burnham, "The Changing Shape of the American Political Universe," *American Political Science Review* 59 (1965): 7–28; Walter Dean Burnham, "Party Systems and the Political Process," in William N. Chambers and Walter Dean Burnham, eds., *The American Party Systems: Stages of Political Development* (New York: Oxford University Press, 1967); Walter Dean Burnham, *Critical Elections and the Mainsprings of American Politics* (New York: W. W. Norton, 1970); Walter Dean Burnham, "Constitutional Moments and Punctuated Equilibria: A Political Scientist Confronts Bruce Ackerman's *We The People*," *Yale Law Journal* 108 (1999): 2237–77; Jerome M. Clubb, William H. Flanigan, and Nancy H. Zingale, *Partisan Realignment: Voters, Parties, and Government in American History* (Beverly Hills, CA: Sage, 1980); James L. Sundquist, *Dynamics of the Party System Alignment and Dealignment of Political Parties in the United States* (Washington, DC: Brookings Institution, 1983).

73. David R. Mayhew, *Electoral Realignments: A Critique of an American Genre* (New Haven, CT: Yale University, 2002).

74. Burnham, "Constitutional Moments."

75. Frank R. Baumgartner and Bryan D. Jones, *Agendas and Instability in American Politics* (Chicago: University of Chicago, 1993), p. 236.

76. Ibid., p. 19.

77. My argument as to the metaphorical nature of punctuated equilibria in realignment theory is similar to the one made by Carl Gans in a professional journal in 1987. My account differs from his in two respects: first, in the relatively greater and more explicit emphasis I place on operationalization of concepts; and second, in my advantage in being able to quote Gould's own reaction to the uses that had been made of his theory. See Carl Gans, "Punctuated Equilibria and Political Science: A Neontological View," *Politics and the Life Sciences* 5, no. 2 (February 1987): 225.

78. Gould, *Structure*, p. 46.

79. Stephen Jay Gould, *I Have Landed: The End of a Beginning in Natural History* (New York: Three Rivers Press, 2003), p. 227; Stephen Jay Gould, *Leonardo's Mountain of Clams and the Diet of Worms* (New York: Three Rivers Press, 1998), p. 298.

80. Gould, *Wonderful Life*, p. 99; see also p. 212.

81. Ibid., p. 319.

82. Ibid., p. 208.

83. Ibid., p. 319.

84. Ibid., p. 236.

85. Ibid., pp. 47, 48, 236, 239.

86. Ibid., p. 233.

87. Ibid., p. 14.

88. Ibid., p. 50.

89. Stephen Jay Gould, *Full House: The Spread of Excellence from Plato to Darwin* (New York: Three Rivers Press, 1996), pp. 181–90.

90. Ibid., pp. 176.

91. Stephen Jay Gould, *The Flamingo's Smile: Reflections in Natural History* (New York: W. W. Norton, 1985), p. 198.

92. Stephen Jay Gould, *Dinosaur in a Haystack: Reflections in Natural History* (New York: Harmony Books, 1995), p. 246.

93. Gould, *Leonardo's Mountain of Clams*, pp. 211–12.

94. Ibid., p. 249.

95. Mayr, *Diversity of Life*, p. 317.

96. George Gaylord Simpson, *This View of Life: The World of an Evolutionist* (New York: Harcourt, Brace, and World, 1964), pp. 21, 23, 265, 267.

97. Timothy Shanahan, *The Evolution of Darwinism: Selection, Adaptation, and Progress in Evolutionary Biology* (Cambridge: Cambridge University Press, 2004), pp. 199–200.

98. Richard Dawkins, *A Devil's Chaplain: Reflections on Hope, Lies, Science, and Love* (Boston: Houghton Mifflin, 2003), pp. 204–205.

99. Dennett, *Darwin's Dangerous Idea*, p. 303; Shanahan, *Evolution of Darwinism*, pp. 207–13; Levinton, *Macroevolution*, pp. 460–68; Edward O. Wilson. *The Diversity of Life* (Cambridge, MA: Harvard University Press, 1992), pp. 183–94.

100. Simon Conway Morris, *The Crucible of Creation: The Burgess Shale and the Rise of Animals* (Oxford: Oxford University, 1998), p. 170.

101. Ibid., pp. 207, 218.

102. Simon Conway Morris, *Life's Solution: Inevitable Humans in a Lonely Universe* (Cambridge: Cambridge University Press, 2003), p. xii.

103. Ibid., p. 116.

104. Ibid., pp. 157–66.

105. Ibid. p. 190.

106. Ibid., pp. 203, 246–53, 258.

107. Ibid., p. 310.

108. Ibid., p. 282.

109. Ibid., p. 321.

110. Ibid., pp. 329–30.

Chapter 4

the politics of human nature

Every great political theory begins, explicitly or implicitly, with assumptions about human nature. For Aristotle in the fourth century BCE, humans, like seeds, contained potential unfolding and growth. Just as a seed sprouts into its potential as a tree, a process that requires the right kind of soil, water, and climate, humans grow into their potential for civilized life if they are planted in the appropriate kind of political community. This community is the quasi-democratic city-state known as the *polis*.[1] For Thomas Hobbes in the seventeenth century, the fundamental situation of human life was physical insecurity. Hence, the primal rule of human behavior is to seek peace and value it above all other things. It follows that only absolute monarchy is natural to the situation of humanity.[2] For John Rawls in the twentieth century, human nature consisted of moral equality among individuals plus the psychological assumption that, when faced with designing a social structure without knowing where they would fit into it, people would create a society based on the "maximin" strategy—maximizing the welfare of the person in the worse-off position. Rawls is not explicit about the political structure deriving from these assumptions, but he hints that it would look like liberal democracy with a very robust welfare state.[3] In each case, philosophy rests upon an understanding of the way nature expresses itself through human personality.

Similarly, when people argue over public policy, they often assert, or imply, that their preferences are somehow consistent with human psychology, whereas their opponents' preferences would, if adopted, require humans to change in fundamental ways. Sometimes they argue that a given policy could never be implemented because human nature would not permit it.

The reverse is also true. People argue that this or that conception of human nature implies consonance with a set of political outcomes—who wins and loses—whereas another conception implies another set of outcomes. Even in scientific discussion, the political implications of a given methodology, or a given theory, are part of the evaluation of the empirical results. In contemporary scientific discourse, leftists are far more explicit in making this type of argument. When leftist scientists criticize someone else's research, they often contend not only that the project is scientifically mistaken, but that its assumptions tend in some way to strengthen existing political inequalities. As Ruth Bleier puts it, there are certain kinds of scientific theories "that implicitly defend the *status quo*."[4] In biology there is a group of politically aware leftist scholars, of whom Gould was one, whose members share this view—some scientific theories have a built-in tendency to reinforce socially constructed dominance hierarchies by making them appear to be the product of nature.

In contrast, the people who are accused by Gould and his allies of promulgating right-wing science by creating mistaken theories of human nature deny the charge. In contemporary scientific debate, no one admits to fostering theories that tend to reinforce the political status quo. Sometimes right-wing journalists pick up a particular theory and use it to justify patriarchy, or sexism, or racism, or capitalism, but when this happens the theory's author almost always disavows the political inference.

The contemporary debate is thus not symmetrical, with one side frankly offering theories that they believe support political equality and the other side offering theories that candidly support established power relationships. Instead, the debate within evolutionary biology over the final three decades of the twentieth century featured a small group of leftists offering their theories, which they insisted supported equality, and a larger group of scientists who insisted only on the nonpolitical purity of their vocation. The leftists analyzed the writings of their opposition in order to demonstrate their tendency to reinforce the status quo, while the members of the opposition accused the leftists of traducing the purity of science by dragging in irrelevant political matters.

The writing of the majority in this argument frequently betrayed much irritation that Gould and his allies were using deception and obfuscation to mislead the public as to the true nature of the majority's research agenda. Even Ullica Segerstrale, who as a sociologist as well as a biochemist might have been expected to show more tolerance of political content, objected to Gould's "moral reading" of science, his "coupled agenda," and his pursuit of "politics by scientific means."[5] Gould was often accused of confusing the public by mixing political irrelevancies with scientific information in his popular essays.[6]

No doubt the conflict was confusing to outside observers, and even to those in the middle of it. But it was not Gould's fault that so much of the argument over human nature was, and is, perplexing. Political disputation is always ambiguous, for reasons articulated very well by political scientist E. E. Schattschneider in 1960:

> Political conflict is not like an intercollegiate debate in which the opponents agree in advance on a definition of the issues. As a matter of fact, *the definition of the alternatives is the supreme instrument of power*; the antagonists can rarely agree on what the issues are because power is involved in the definition. . . . It follows that all conflict is confusing.[7]

When Gould's opponents argued that they were pure scientists attempting to pursue a disinterested agenda, while he was attempting to introduce irrelevant considerations, they were themselves making a political statement. They were attempting to define the conflict so that the public would see them as above politics. As any observer of American politics knows, defining oneself as a public-spirited citizen and one's adversaries as politically motivated is a time-worn tactic in democratic maneuvering.

Doubtless, there are some areas of science in which the data are unambiguous and the interpretations are clean cut; these areas, primarily in physics and chemistry, offer little room for interpretation. Some biological evidence, especially at the genetic level, is likewise unequivocal and does not support political controversy. There is no argument, for example, over the genetic etiology of sickle-cell anemia. But a wide range of the evidence in evolutionary biology—and this is particularly true of paleontology—is often fragmentary and problematic, lending itself to arguments over interpretation. It invites ideological readings, and is thus more like evidence in the social sciences. When Gould's opponents accused him and his allies of importing tendentious irrelevancies into biology, therefore, they were either making their own political statement or advertising their naiveté. In evolutionary biology, politics comes with the territory.

The individuals who participated in the controversies within the scientific study of evolution during the years from 1972 to 2002 were as brilliant, independent minded, and cantankerous as any other group of professional intellectuals. They always insisted on expressing their unique points of view; they did not run in packs. Gould did not agree, on every topic, with his sometime coauthors Niles Eldredge and Richard Lewontin. Edward O. Wilson and Richard Dawkins, often characterized together as "ultra-Darwinists" by Gould and his friends, expressed opinions that diverged at many important points. Ernst Mayr, perhaps the most eminent evolutionist of the twentieth century, strongly agreed with Gould on some topics and strongly disagreed with him on others. Any generalization is bound to distort the views of somebody. Yet it is not misleading of me to speak of two intellectual sides or teams that contended within evolutionary biology during that era, if only because the arguers themselves recognized the dichotomy. When Dawkins wrote, he praised John Maynard Smith, Robert Trivers, and Wilson, and criticized Gould, Lewontin, and Eldredge. When Eldredge published, he reversed the targets of praise and criticism. I may err in summarizing the beliefs of the members of each team, but the teams existed independently of my description.

To the leftists, human nature was almost infinitely malleable. This great latitude in behavior did not come about because *Homo sapiens* had somehow escaped the logic of natural selection. Rather, among humans the genetic heritage had led to a species whose essence was social. "Human nature" in any given society thus depended on the cultural context, meaning that historically common cultural patterns—racism, say, or patriarchy—could be erased given the right cultural forces. As Richard Lewontin put it, having made the brain,

> the genes have made possible human nature, a social nature whose limitations and possible shapes we do not know except insofar as we know what human consciousness has already made possible. . . . The genes, in making possible the development of human consciousness, have surrendered their power both to determine the individual and its environment. They have been replaced by an entirely new level of causation, that of social interaction with its own laws and its own nature that can be understood and explored only through that unique form of experience, social action.[8]

Leftist scientists prefer an ontology, methodology, and epistemology that they believe emphasizes cultural independence from genetic universals. As Lewontin said to me in 2006, "The world must be malleable; human nature cannot be fixed. A reformist person must believe that changing people is possible."[9] On the left, it is imperative that social arrangements be viewed as *not natural*, as artificial contrivances that can be undone. Similarly, it is convenient for right-wingers to believe that existing social arrangements are biological givens and that it would be futile to attempt to reconfigure them. A right-wing scientist (or citizen observing science) therefore is apt to employ an ontology, methodology, and epistemology that assumes cultural dependence on genetic universals.

It is not my contention that Gould, or Lewontin, or anyone else, realized their political values when young, then set out to fashion a set of scientific strategies that would produce the appropriate political conclusions. Neither on the left nor on the quasi-right are intellectuals so crude. The essential point about the political views of scientists is that the science and the politics grow naturally together as different facets of the same worldview. It is tempting to employ some sort of organic metaphor here, comparing different aspects of a personal ideology to the separate organs of a body, outwardly diverse in form and function but growing from the same primal set of instructions and cooperating to the same ultimate end. But such a metaphor would be misleading, because while the relationship of organs in a body is identical in every organism of a species, the scientific and political ideas in every scientist relate in distinct ways. When I discuss Gould's interpenetration of scientific and political ideas, it cannot be assumed that the same relationships apply to all leftists. Gould generally comes to the same conclusions as other leftists, but not necessarily through the same pattern of reasoning.

Among Gould's opponents, who would vehemently deny that they are right wing, the master assumption is that there is a human nature, and that it is the motive force for all the variations of culture. As psychologist Steven Pinker puts it,

> Whatever the exact picture turns out to be, a universal complex human nature would be part of it. I think we have reason to believe that the mind is equipped with a battery of emotions, drives, and faculties for reasoning and communicating, and that they have a common logic across cultures, are difficult to erase, or redesign from scratch, were shaped by natural selection acting over the course of human evolution, and owe some of their basic design (and some of their variation) to information in the genome.[10]

And Edward O. Wilson, in a sentence that is often quoted by leftists, summarizes his position by writing that "the genes hold culture on a leash."[11]

It goes without saying that, in Wilson's view, the leash is relatively short—culture is so closely tied to genetics that behavioral variations can be deduced from evolutionary theory. Gould argues, in contrast, that "the leash is loose and nonconstraining," so much so that evolutionary theory "cannot specify a particular institution but only a broad range of possibilities."[12]

This is not a dispute over "nature" versus "nurture," but different emphases on the degree to which culture is independent of genetic influences. Gould, Lewontin, Bleier, and their allies do not argue that there is no human nature, only that it allows for so much cultural variation that evolutionary theory is in practical terms useless for discussing human behavior. Wilson, Pinker, and their allies would not argue that evolutionary theory explains all variation in cultural practices. Instead, as John Alcock puts it, "Every behavioral trait depends on evolved physiological systems whose proximate development requires both genetic and environmental inputs," a situation that leads to "conditional strategies" of behavior, varying with different environments, that can be studied for all animals, including humans.[13] Some behaviors might be explained using this strategy; some might resist such explanation. But in general, the leash is short enough that many important advances can be made in explaining cultural variations.

These are the global, overall ideological differences that sort biologists into contending political factions. But in the scientific literature, the big ideologies only rarely confront each other directly. Instead, the adversaries tend to fight out their political arguments at the level of scientific methodology.

Adaptationism

When evolutionary biologists embark on the project of explaining something about an organism—its morphology, its behavior, its survival, or its extinction—they have, broadly speaking, two traditions upon which to draw. The first, dominant in Britain and the United States ever since Darwin, is called the *adaptationist program*. The scientist asks, "What is the function of this organ?" or, "What does it do to permit the organism to survive?" The scientist then posits some tentative functional hypotheses that might explain the existence of the organ, tests the hypotheses, and proceeds according to the outcome of the tests. Thus, when Harvey asked himself, "What is the purpose of valves in the veins of vertebrates?" he was led to discover the circulation of blood.[14] When Bates asked, "Why do some animals closely mimic other animals?" he discovered first that the mimicked animal tended to be poisonous, and second, that the geographic distribution of the mimics closely paralleled that of their models.[15] When von Frisch asked, "Why are flowers so colorful?" he was inspired to discover, in opposition to the beliefs of his day, that honeybees have color vision.[16] As philosopher Michael Ruse puts it, "It is because things do work so well so much of the time that we can feel justified in the adaptationist paradigm and can so profitably seek out and find explanations for the exceptions."[17]

The other tradition of inquiry, called *formalism*, has tended to dominate on the continent of Europe, but has always had adherents in the Anglo-American countries. In this tradition, scientists emphasize the structural constraints on organismal development. Once launched with general phyletic tendencies—say, with an internal as opposed to an external skeleton—a sequence of organisms can only develop in a certain direction. No matter what the environmental conditions, lobsters will never evolve into mice, or vice-versa, because their evolutionary pathways have forever constrained their forms and behaviors. As Gould summarized the distinction, "Adaptationists hold that structures must evolve or be fashioned for utility; functional needs come first, and form follows. Formalists argue, on the other hand, that . . . form comes first, and organisms may then discover usages."[18]

Without actually rejecting adaptationism, Gould argued that it was given far too much emphasis in Anglo-American evolutionary biology. He frequently urged his compatriots to pay more attention to Continental biology. Natural selection, and therefore adaptation, were always operating, but there was an unknown percentage of cases in which "constraints restrict possible paths and modes of change so strongly that the constraints themselves become much the most interesting aspect of evolution."[19] Gould wanted to shift the emphasis toward nonadaptationist explanations, to right the balance.

Taking his own advice, Gould often used his essays to suggest a formalist approach to the explanation for some feature of an organism. As one example among many, he addressed the antlers of the Irish elk, the monstrous head ornaments of a huge, extinct ungulate. Known by its fossils for hundreds of years, the Irish elk has long been a topic of evolutionary debate. Adaptationists asked the question, "What could those giant antlers have been good for?" and proceeded to come up with hypotheses that either stressed the adaptive value of the antlers in fighting off other males of the species during the rut, or in appealing to females. Gould, however, showed that the size of the antlers could be explained with reference to the concept of *allometry*—the tendency of some body parts to increase in size more rapidly than others. It happens that as male members of different deer species become larger, their antler size increases faster than their body size. If, as Gould pointed out, increased body size was favored by natural selection, then correspondingly out-of-proportion antlers would have been a by-product of the increase in the species' size. No adaptationist explanation was necessary.[20]

The frequency with which he wrote about such examples of organisms to which formalist explanations might apply is evidence enough that Gould found nonadaptationism sufficiently interesting on a scientific plane. But, as with so much else in his thinking, the deemphasis of the adaptationist program had a political purpose, also. "The contingency of history [which I have discussed in chapter 3] and human free will," he wrote in a 1993 discussion of adaptationism, "are conjoined concepts." The context of his treatment leaves no doubt that adaptationism was another concept essential to the topic.[21] The free will he spoke of is cultural, not necessarily individual. The more formalist, nonadaptationist explanations dominated biology, the more freedom cultures had to permit creativity in human behavior patterns. The more adaptationist explanations dominated, the less cultural variation the genes were going to permit humans to create.

The adaptationist program is explicitly a scientific agenda for allowing scientists to predict behavior. As Edward O. Wilson, the leader of the movement to apply Darwinist insights to human behavior, summarizes his methodology,

> To dissect a phenomenon into its elements . . . is consilience by reduction. To reconstitute it, and especially to predict with knowledge gained by reduction how nature assembled it in the first place, is consilience by synthesis. That is the two-step procedure by which natural scientists generally work. . . . The central idea of the consilience world view is that all tangible phenomena, from the birth of stars to the workings of social institutions, are based on material processes that are ultimately reducible, however long and tortuous the sequences, to the laws of physics.[22]
>
> The power of a scientific theory is measured by its ability to transform a small number of axiomatic ideas into detailed predictions of observable phenomena.[23]

As I discussed in chapter 2, prediction implies determinism. If a scientist is able to predict the behavior of a bluebird under given circumstances, then the behavior of that bird is determined under those circumstances. This conclusion is equally true whether the prediction is absolutely precise or probabilistic. It follows that if scientists are able to predict human behavior, that behavior is also determined. Gould's fundamental political objection to the adaptationist program seemed to be that by restricting its vision to the predictable it assumed that human behavior is not free, within culture, to vary markedly, but is held on a short leash by the genes.

In thus worrying about the political implications that ride along with adaptationism, Gould was somewhat less consistent than his chief collaborator in his writings on this subject, Richard Lewontin. In several publications, Lewontin expressed the opinion that adaptationism should be jettisoned entirely as a scientific approach to the study of life:

> The metaphor of adaptation, while once an important heuristic for building evolutionary theory, is now an impediment to a real understanding of the evolutionary process and needs to be replaced altogether. Although all metaphors are dangerous, the actual process of evolution seems best captured by the process of *construction*.[24]

Indeed, if he believed that adaptationism really did have reprehensible political implications, then Gould should have followed Lewontin in rejecting the entire approach. If he thought that the adaptationist program fostered nefarious tendencies, then he should have tolerated none of it. Instead, throughout his writings he pulled back from complete consistency, always finding fault with adaptationism as employed by others but never rejecting the methodology completely. Even in his last book he asserted, "I seek a fusion of structural and functional influences," rather than clearly embracing the structural formalism that would seem to have been his logical position.[25]

Be that as it may, the attack on the adaptationist program gave Gould an opportunity for his second great goosing of the Darwinian orthodoxy. In late 1978 Richard Lewontin received an invitation to present a paper at a conference on adaptationism to be held at the Royal Society in London. Lewontin was unable to attend, so he asked Gould, his friend, colleague, and ideological consociate, to compose the paper and deliver it at the conference. As

Gould later wrote of the final product, "The ideas are more his, the writing almost entirely mine."[26] His presentation caused another explosion within evolutionary biology, one that still reverberates.

Once again, Gould's compelling rhetorical style, and especially his facility with metaphor, were at least as important in stimulating the reception of this article as were the ideas it contained. It was so vivid and over the top in style that it stimulated analyses by professors of rhetoric. In a collection of essays published in 1993, various scholars pointed out that Gould created "sharp divisions" within biology where before there had been ambiguity, in pursuit of his creation of "an intertextual drama" in which the adaptationists were the villains.[27] He was described as having "remained consistently civil while advancing a strenuous polemic," a characterization half of which the targets of his attack would dispute.[28] It was pointed out that Gould had recast "a disciplinary issue as essentially a moral one," which meant, of course, that those whom he had tarred as immoral would react with moral outrage.[29] The same scholar also noticed that he used shorter sentences and a more informal vocabulary when lampooning adaptationism, thus seeming to "appear at the reader's elbow" so that "together author and reader share the joke." He shifted to longer sentences and a more formal vocabulary when outlining his alternative, thus establishing his seriousness as an authentic scientist.[30]

This essay was available for analysis by people outside of a natural science discipline because, as with much else of Gould's writing, it was accessible to intelligent people who had no special training in biology. Although some of the essay was a dense analysis of the work of the German paleontologist Adolf Seilacher, whose research Gould and Lewontin believed had been unfairly neglected, most of it was an entertaining philippic.

Gould opened with one of his favorite nonbiological topics, the architecture of medieval churches, fashioning a metaphor that still rings in the heads of biologists. The central dome of St. Mark's cathedral in Venice, he related, is constructed of four concave arches, meeting at the apex. On the inside of the dome, the undersides of the arches are decorated with religious symbolism. The quadrants alone, suitably constructed, are sufficient to bear the load of the domed ceiling. But curving the quadrants toward the apex leaves curve-sided triangular spaces between each quadrant unfilled. To keep the rain out, the architect has completed the dome by covering the openings with concave spandrels—architectural additions that are unnecessary to the load-bearing structure of a building. The architect has decorated the underside of the spandrels as gloriously as the underside of the load-bearing quadrants, and as a result, from inside the cathedral, the spandrels are indistinguishable, in an aesthetic sense, from the quadrants.

Gould pointed out that the spandrels "are a necessary architectural by-products of mounting a dome on rounded arches."[31] But the artistic power of the whole underside of the dome is such that it can fool the architecturally uneducated into thinking that the spandrels are as important as the quadrants. "The design is so elaborate, harmonious and purposeful that we are tempted to view it as the starting point of any analysis, as the cause in some sense of the surrounding architecture. But this would invert the proper path of analysis. The system begins with an architectural constraint."[32] A proper understanding of the cathedral begins with the insight that the arches are essential, while the spandrels are, in a manner of speaking, merely along for the ride.

Among tourists the misunderstanding of the role of the spandrels is of no importance. But, said Gould, biologists often made an analogous mistake when attempting to explain the features of organisms, and this error was deeply significant.

Many features of organisms are simply by-products of their structural development, and may have no adaptive significance. Obvious examples would be the redness of blood and the whiteness of bones, which bear no relationship to the adaptive value of those organs. As another example, all snails that grow their shells by coiling around an axis must generate a small cylindrical space, termed an *umbilicus*, at the center of the coiling. Some snail species use the umbilicus as a brood chamber to protect their eggs, but it would be a mistake to portray the umbilicus as an adaptation to the need for egg protection, that is, as having been evolved in response to the necessity for such protection. The umbilicus is a spandrel of the coiling shell; only secondarily did it become adaptively useful.[33]

Such examples could be expanded to an unknown extent. Yet "evolutionary biologists, in their tendency to focus exclusively on immediate adaptation to local conditions, do tend to ignore architectural constraints and perform just such an inversion of explanation."[34] The adaptationist program "regards natural selection as so powerful and the constraints upon it so few that direct production of adaptation through its operation becomes the primary cause of nearly all organic form, function, and behavior. . . . We maintain that alternatives to selection for best overall design have generally been relegated to unimportance by this mode of argument."[35]

Having created a vivid metaphor to make his point, Gould then seemed to slip into rhetorical overdrive, fashioning a second metaphor that was equally memorable but far less useful to science because it was deliberately insulting to the majority of professionals in his and related fields. The adaptationist program, he asserted, should actually be termed the "Panglossian paradigm," after Voltaire's fictional doctor who proclaimed that this was the best of all possible worlds. Most evolutionary biologists see every organismic trait as tending toward perfection, because the adaptationist worldview emphasizes "the near omnipotence of natural selection in forging organic design and fashioning the best among possible worlds."[36] When encountering a puzzling organic form or behavior, biologists habitually assume that the object of study must represent an optimal adaptation to an evolutionary problem. They then search for the problem to which the form is an optimal solution. In aggregate, the view of evolution that biologists create is one that suggests that "our world may not be good in an abstract sense, but it is the very best we could have. Each trait plays its part and must be as it is."[37]

It would be hard to exaggerate the extent of negative emotion, and, more positively, the depth of soul-searching, sparked by this article. Gould had written, at various points in the exposition, that he was merely arguing for a reorientation of thinking to take more account of possibly nonadaptive explanations. His argument "does not deny that change, when it occurs, may be mediated by natural selection," but that there may also be other forces at work, and that the relative importance of natural selection and the other forces must be discovered through empirical investigation.[38] But the ridicule of his rhetoric was so stinging that other biologists perceived it as an attack on all adaptationism at all times, and, incidentally, on them personally. The fact that the complete rejection of adaptationism, may, in fact, have been Lewontin's position, as developed in subsequent publications, strengthened the tendency of biologists and their allies to ignore Gould's nuances and circle the wagons.

The return fire continued for years.[39] Gould's attack on the adaptationist program was simply "unjustified teleological panglossianism," wrote Tooby and Cosmides.[40] Jeffrey Levinton opined, "The extremists [Gould and Lewontin] would have us believe that nonadaptive aspects of form have been woefully neglected. . . . This point of view is a misunderstanding of typical biological practice. It is true that when seeing a structure for the first time, nearly all biologists ask 'What does this do?' . . . Such an approach . . . has been . . . universally successful."[41] "Adaptationist reasoning is not optional; it is the heart and soul of evolutionary biology," lectured Daniel Dennett. If Gould "really wants to ask and answer 'why' questions, he has no choice but to be an adaptationist."[42]

By the time of Gould's death in 2002, it was clear that he and Lewontin had not reoriented evolutionary biology away from the adaptationist program. Nevertheless, they had had a major impact on the mindset of biologists, although Tooby, Cosmides, Levinton, and Dennett would probably prefer not to admit it. Both Gould and Lewontin had continued to write books and articles criticizing adaptationism, and their initial attack plus its sequels stimulated an era of "Post-Spandrel Adaptationism" in evolutionary biology.[43] Biologists were driven to forestall the charge of being blind adaptationists by inventing better methods of identifying and measuring adaptationism when it occurred. One technique was to turn a hypothesis into a formal mathematical model that would then be used to derive operational predictions. Another was to develop methods of comparing a given species with its relatives, in order to be able to distinguish adaptations specific to that species from traits that were common to all the related species.[44] Even when not using new methodologies, researchers were now quite aware of the need to explore alternative explanations to adaptationism. Thus, while biologists remained adaptationists, their thinking was powerfully affected by the desire to avoid making themselves vulnerable to the charge of being a Pangloss.

At the level of biology, stripped of rhetoric and politics, Gould continued for the rest of his career to analyze examples of structures that, he maintained, were spandrels, and thus could not be explained within the adaptationist program. The outstanding examples of such evolutionary spandrels, he argued several times, were the various remarkable capacities of the human brain. He pointed out that Alfred Russel Wallace, the cocreator of the theory of natural selection, had given up the attempt to understand brain capacity as an adaptation. The people in the primitive tribes with whom Russel had lived early in his career had, as nearly as he could tell, the same intellectual range as Europeans. That is, they would have been capable of attending an opera or learning physics in college. Yet they lived in a culture in which such activities were never thought of. As Gould recounted Wallace's reasoning,

> If natural selection constructs organs for immediate use, and if brains of all people are equal, how could natural selection have built the original "savage's" brain (his terminology)? After all, savages have capacities equal to ours, but they do not use these abilities in devising their cultures. Therefore, natural selection, a force that constructs only for immediate utility, cannot have fashioned the human brain.[45]

Unable to imagine how his theory could account for the abilities of the brain, Wallace finally abandoned it and embraced religion—God created the human brain, and natural selection created everything else. But, said Gould, if Wallace had only had the concept of the spandrel at his disposal, he would not have despaired of explaining his own mind. *Most* of the abilities of the brain are spandrels, by-products of an organ that evolved to deal with other problems:

> Our ability to read and write has acted as a prime mover of contemporary culture. But no one could argue that natural selection acted to enlarge our brains for this purpose. . . . Selection made our brains for other reasons, while reading and writing rose later as a fortuitous or unintended result of an enlarged mental power directly evolved for different functions.[46]

If one takes the idea of spandrels seriously, however, one has to come to the further thought that some features—not the redness of blood, but the use of the snail's umbilicus as a brooding chamber—that are accidental by-products of some adaptive organ are themselves sometimes assimilated for adaptive use. Such an assimilation, Gould argued, would be the computational capacities of the brain. In order to fix this concept more firmly in the awareness of the discipline, Gould and Elisabeth Vrba invented the term *exaptation*, defined formally as "characters, evolved for other usages (or for no function at all), and later 'co-opted' for their current role."[47] For the rest of his career, Gould argued that the brain of *Homo sapiens* was more important for its exaptations than its adaptations. As a consequence, he asked rhetorically, "Mustn't the ever-cascading spandrels of the human brain be more weighty than the putative primary adaptations of ancient African hunter-gatherer ancestors in setting the outlines of what we now call 'human nature?'"[48]

The answer to that question illustrates the way Gould and Lewontin folded their attack on the adaptationist program into another, even more intense conflict. For adaptationism was only one front in a war that had begun in biology in the mid-1970s, and then spilled out into the larger society.

Sociobiology

The war over sociobiology actually started four years earlier than the presentation of Gould and Lewontin's famous polemic against the adaptationist program. I have discussed their attack on adaptationism first, however, because it is logically prior to their opposition to sociobiology. Although their rationale for rejecting adaptationism was not published until 1979, it clearly underlay the assault on sociobiology that dated to 1975.

The conflict began with the publication of Edward O. Wilson's *Sociobiology*.[49] Although there had been a vast amount of evolutionary biology that included not just the form but the behavior of animals, it had been scattered over a variety of subfields, and its practitioners had no self-awareness that they might be considered to be working on the same subject matter. Wilson, however, adopted the term *sociobiology*, which had been in use without a stable meaning since the 1940s, and defined it as "the systematic study of the biological basis of all social behavior," thus bringing self-consciousness to a community of scholars.[50]

Although he himself did not at first recognize the fact, Wilson's ontology was bound to raise the ire of Gould and Lewontin, his two colleagues in the Harvard biology department. First, his approach to the study of behavior through evolutionary biology was reductionist. "In a Darwinian sense, the organism does not live for itself," he wrote. "Its primary function is not even to reproduce other organisms; it reproduces genes, and it serves as their temporary carrier."[51] As I have discussed in chapter 2, this methodological stance alone was enough to make leftist scientists nervous. Second, however, and even more alarming to the left, the sociobiological ideal was frankly predictive. "The principal goal of a general theory of sociobiology should be an ability to predict features of social organization from a knowledge of . . . population parameters combined with information on the behavioral constraints imposed by the genetic constitution of the species,"—a clear statement from the first chapter.[52] Third, as a companion to prediction, sociobiology relied upon the adaptationist program, recommending a methodology in which "each phenomenon is weighed for its adaptive significance and then related to the basic principles of population genetics."[53] And fourth, the final chapter of *Sociobiology* was an attempt to sketch out a research agenda for applying the same reductionist, predictive, adaptationist science to the study of *Homo sapiens*. "There is a need for a discipline of anthropological genetics," Wilson proclaimed. " We are compelled to drive toward total knowledge, right down to the levels of the neuron and the gene. When we have progressed enough to explain ourselves in these mechanistic terms, and the social sciences come to full flower, the result might be hard to accept."[54]

The subfield of sociobiology immediately became a smash hit in natural science, and has only grown in influence in the years since publication of Wilson's book. As Wilson reported in his autobiography, in the twenty years after 1975 more than two hundred books were published on human sociobiology and related topics, and his initial offering inspired the founding of four new journals.[55] In particular, the field of evolutionary psychology, one of the fastest growing and most prestigious areas in the social sciences, owes its inspiration to *Sociobiology*.[56]

But for all the success of sociobiology among scholars, Wilson was more a prophet than he knew when he predicted that some of the consequences of his new field might be "hard to accept." A science that was reductive, predictive, adaptationist, and specifically aimed at explaining human behavior was the Platonic ideal of the leftist scientific nightmare. Reductionism, in their view, implicitly endorsed the false notion that humans are individuals rather than members of a social whole. Predictionism and adaptationism, by assuming that genes hold culture on a short leash, denied to humans the ability to construct their own patterns of behavior, and therefore left them subject to historically dominant modes of unjust domination. Implicitly, therefore, sociobiology endorsed, by acceptance, existing institutions of inequality as natural—as written in the genes through evolution. Although Wilson convincingly denied, then and afterward, that his intentions were political—or even that his thinking was anything but strictly scientific—his creation of the subfield of sociobiology could not but provoke an intense reaction from his colleagues on the left.

Shortly after *Sociobiology* was published to generally rapturous reviews, Gould, Lewontin, and a group of other scientists and graduate students wrote a broadside in the *New York Review of Books* that set off a clamorous controversy within the intellectual community and the hyperpoliticized subcultures of Cambridge and New York. Sociobiologists, they charged, belonged to "that long parade of biological determinists whose work has served to buttress the institutions of our society by exonerating them from responsibility for social problems."[57] They lumped Wilson together with a motley group of leftist boogeymen, including Social Darwinism, proponents of innate racial difference in intelligence, restrictive US immigration laws, and Nazi gas chambers, suggesting that "the reason for the survival of these recurrent determinist theories is that they consistently tend to provide a genetic justification of the *status* quo and of existing privileges for certain groups according to class, race, or sex."[58] Wilson responded with injured innocence, and various other participants jumped in, then and for years afterward.

The history of this loud and colorful conflict has been well chronicled in a variety of publications, especially Ullica Segerstrale's magisterial *Defenders of the Truth.*[59] The scholarship that has already been produced establishes beyond doubt that the polemics and counter-polemics were full of enough overstatement, slander, misrepresentation, misunderstanding, irrelevancy, and invective (on both sides) to satisfy any connoisseur of human misbehavior. I intend not to add to the total of general comment, but only to discuss some particular aspects of the controversy that I think have perhaps not been viewed from quite the correct angle.

The first is the argument over whether practitioners of sociobiology were advocating "genetic determinism" as a mode of inquiry. For years after 1975 Gould and Lewontin maintained, in many different publications, that the whole field was ineluctably determinist. As Gould summarized in *Hen's Teeth and Horses Toes* in 1983:

> Sociobiology is not just any statement that biology, genetics, and evolutionary theory have something to do with human behavior. Sociobiology is a specific theory about the nature of genetic and evolutionary input into human behavior. It rests upon the view that natural selection is a virtually omnipotent architect, constructing organisms part by part as best solutions to problems of life in local environments. . . . Applied to humans, it must view *specific* behaviors (not just general potentials) as adaptations built by natural selection and rooted in genetic determinism.[60]

This sort of accusation greatly annoyed sociobiologists. In the second edition of *On Human Nature* Wilson reflected wonderingly, "In the popular media," sociobiology had come to mean "the theory that human behavior is determined by genes, or at least strongly influenced by them, as opposed to learning."[61] John Alcock fumed that "the myth of the determinist sociobiologist has been carried forward by some opponents who avoid acknowledging even in passing the long history of rebuttal to this caricature. Why? Because the genetic determinist is too convenient a strawman to be discarded."[62] And Steven Pinker sniffed that "a given gene may not have the *same* effect in all environments, nor the same effect in all genomes, but it has to have an *average* effect. That average is what natural selection selects (all things being equal), and that is what the 'for' means in 'a gene for X.' It is hard to believe that Gould and Lewontin, who are evolutionary biologists, could literally have been confused by this usage, but if they were, it would explain twenty-five years of pointless attacks."[63]

These injured defenses of sociobiology seem to reflect a certain tone deafness to political thinking on the part of many natural scientists. Wilson, Alcock, and Pinker understood that Gould and Lewontin had political intentions (Pinker even titled his chapter about the controversy "Political Scientists"), but they could not seem to cipher a legitimate connection between political and scientific values. If the intention of sociobiology was the acknowledged one of predicting behavior, however, including human behavior, then its practitioners were in no position to deny that they were genetic determinists. To repeat a quotation from physicist Steven Weinberg, "Determinism is logically distinct from reductionism, but the two doctrines tend to go together because the reductionist goal of explanation is tied in with the determinist idea of prediction: we test our explanations by their power to make successful predictions."[64] If sociobiologists had adopted the debating strategy of maintaining that Gould and Lewontin were actually *antiscience*, because science cannot proceed without being reductionist and predictionist, and therefore, inescapably, determinist, at least they would have been consistent. Instead, they tried to slip away from the accusation by denying that they were determinist, which was, in essence, to deny their own status within the type of science they practiced.

Part of the intellectual turmoil of the sociobiological controversy, therefore, was caused by the puzzling inability of the sociobiologists to meet their opposition on the fair ground of the philosophy of science, and forthrightly defend their own ontological choices. I believe that this inability was caused by the unwillingness of many scientists to acknowledge the political implications of every ontological stance. Gould and Lewontin, by seeing the political implication of each scientific strategy, were at liberty to consider the political consequences of every such choice. The sociobiologists, clinging to an ideal of science that denies its political dimension, were hobbled in scientific polemic. They could accuse their leftist critics of having sullied a virginal science with politics, but they could not engage the issue on its merits.

In Schattschneider's terms, Gould, Lewontin, and their allies had a much broader definition of the issue than the sociobiologists. The leftists refused to narrow their definition to science divorced from politics. The sociobiologists refused to broaden their definition to include the political implications of ontological choices. As Schattschneider would have predicted, the result was confusion.

As it happens, sociobiology, and its offshoots such as evolutionary psychology, have produced a large number of successful predictions about human behavior, and have thereby added considerably to our knowledge of ourselves. Indeed, Pinker's book *The Blank Slate* and Alcock's book *The Triumph of Sociobiology* are convincing compendiums of the knowledge about human behavior that science now possesses because of the adoption of sociobiological techniques. Given this undeniable success, it is curious that sociobiologists continue to deny that they do what they do in the way they do it. I await the sociobiologist who proclaims, "Yes! I am a determinist! And I know more than the nondeterminists ever will!"

The second issue of interest in the sociobiological controversy consists of the conjoined problems of overgeneralization of argument and undergeneration of supporting evidence. My own observation is that sociobiologists, while they can be careful and precise when discussing the application of the theory of natural selection to animals, tend to wax loose and sloppy when discussing humans. In addition, often the generalizations they make tend, just as Gould charged, to reinforce existing social arrangements. Sociobiologists are often sexist and occasionally racist, although statements supporting capitalism are hard to find in the literature. More generally, combing through sociobiological publications yields a collection of quotations that does not seem to fall under the normal expectations of scientific reserve:

> There is, I wish to suggest, a strong possibility that homosexuality is normal in a biological sense, that it is a distinctive beneficent behavior that evolved as an important element of early human social organization.

> Homosexuals may be the genetic carriers of some of mankind's rare altruistic impulses (E. O. Wilson, *On Human Nature*).[65]

> The most important reason for the rarity of democracy is that evolution has endowed our species, as it has with other social primates, with a predisposition for hierarchically structured social and political systems (Albert Somit and Steven Peterson, *Darwinism, Dominance, and Democracy*).[66]

> It seems that the very appearance of full sexual females in position of political power disrupts the male-male, hunting-warring-politicking system. We can accept—even adore—a Jacqueline Kennedy as the desirable wife of one of our leaders, but it may be quite a long time before people of either gender can overcome their biology and accept an overtly sexual woman as a leader in her own right (David Barash, *The Whisperings Within*).[67] [This book was published in 1979, the year Margaret Thatcher became prime minister of Great Britain, and after Golda Meir had led Israel and Indira Gandhi had led India.]

> For males, during the hunter-gatherer period of human evolution, the optimal combination of mating effort and paternal investment varied with the severity of the winters. In Africa, a strong sex drive, aggression, dominance seeking, impulsivity, low anxiety, sociability, extraversion, and a morphology and muscle enzyme suitable for fighting led to male success, whereas in northeast Asia, altruism, empathy, behavioral restraint, and a long life assisted success in provisioning (J. Phillippe Rushton, *Race, Evolution, and Behavior*).[68]

> The processes by which voters decide which candidate information to evaluate, how much candidate information to acquire, and which cognitive mechanisms to use to process the available information, can all be best explained via evolutionary psychology (Gad Saad, "Evolution and Political Marketing").[69]

> This book documents the universal preferences that men and women display for particular characteristics in a mate (David Buss, *The Evolution of Desire*).[70]

The last-named book is a good case study of the way sociobiologists often employ evidence in a rather easy-going manner to make their points. Buss's portraits of male and female mating strategies are not crude, but they do sketch two clear general tendencies, based upon "underlying psychological mechanisms" that are "the products of evolution."[71] While he lists and explores many variations and subtleties in human relationships, his overall picture of love and sex is determinedly orthodox. His conclusions are straightforward deductions from the fact that "males are defined as the ones with the small sex cells [sperm], females as the ones with the large sex cells [eggs]."[72] For a woman an egg is a big metabolic investment. She must therefore not waste her opportunities to mate—she must be choosy, and think about the possible long-term consequences of any given mating. For a man, however, sperm is cheap. Consequently, after mating, he need not hang around to see if anything useful comes of it. A better strategy for him is to attempt to impregnate as many females as possible, in order to spread around his sperm to as great an extent as possible. What we learn from evolutionary theory, therefore, is that men want sex and women want resources; all else is strategy and hypocrisy.

Buss gathers a variety of evidence to support his two dominant portrayals of the male and female characters. He tells us that he surveyed 10,047 persons worldwide, employing fifty collaborators from thirty-seven cultures on six continents and five islands, "from Australia to Zambia."[73] He employs a variety of survey instruments to probe the psyches of his informants, and also brings in a great deal of research, suitably interpreted, from cultural anthropology.

Taken as a whole, the weight of his evidence is impressive, and, since it conforms to pop-culture stereotypes of male and female, without further thought it could be convincing. But when a skeptic begins to probe the evidence itself, as opposed to his *description* of the evidence, its solidity begins to melt away. At one point, for example, he tells us:

> Lesbians tend to be similar to heterosexual women in placing little emphasis on physical appearance, with only 19.5 percent of the heterosexual women and 18 percent of the lesbians mentioning this quality. In contrast, 48 percent of heterosexual men and 29 percent of homosexual men state that they are seeking attractive partners.[74]

Granted, these figures demonstrate that women tend to be less appearance oriented than men. But they also tell us that 52 percent of heterosexual men—that is, a majority—are *not* emphasizing the attractiveness of their partners. What evolutionary theory predicts that only a minority of men will be interested in physical attractiveness?

Similarly, in buttressing the claim that men are sex oriented whereas women are not, Buss reports, "Fantasies about group sex occur among 33 percent of the men but only 18 percent of the women."[75] Again, while the survey results do show that women are less sex oriented, they also show that *two-thirds* of the men are *not* sex oriented by this measure.

Additionally, when supporting the argument that men are ever ready to seduce women, Buss reports that "when the women were asked whether a man had ever deceived them by his exaggeration of the depth of his feelings in order to have sex with her, 97 percent admitted that they had experienced this tactic at the hands of men; in contrast, only 59 percent of the men had experienced this tactic at the hands of women."[76] Again, if we assume that both sexes are being equally truthful here, then an overwhelming majority of women had experienced deception, but a large majority of men had also. An evolutionary theory that posits different behaviors in the sexes based on the size of their sex cells is disconfirmed, not supported, if majorities of both sexes behave the same way.

Some of Buss's data do seem to support his theory. But his heavy reliance on the sort of very problematic evidence discussed above must lead to the general impression that he is overgeneralizing from an unsatisfactory and ambiguous foundation. Given the fact that his book largely reinforces sexual stereotypes, it seems to be an example illustrating Gould's assertion that much of sociobiological research tends to encourage people to suspect "that their social prejudices are scientific facts after all."[77]

This is only one example of the impeachable use of evidence in the sociobiological tradition. In his book *Vaulting Ambition*, however, philosopher Philip Kitcher examined the incaution of many human-oriented sociobiologists in detail, demonstrating the way their conclusions frequently outran the available evidence.[78] It seems to be a common failing in the genre.

The important question is to decide whether the evident tendency of such research to reinforce various social stereotypes is, as Gould, Lewontin and their allies charged, an inevitable consequence of sociobiological ontology and method. The general indictment is that by employing the adaptationist program to study *Homo sapiens*, sociobiologists must, without exception, produce work that is supportive of social inequalities of race, class, and sex. They must see whatever cultural patterns that exist, assume that they have been adaptive over evolutionary time, and conclude that they are therefore not only permanent, but good.

I find the general argument that conservative political patterns are inherent in the adaptationist program to be unconvincing. In a previous section of this chapter I tried to establish that while Gould's rhetoric in the 1979 article with Lewontin was uncompromisingly hostile to adaptationism, over the course of his career he showed that he was not actually opposed to its use, as long as that use was leavened with structural approaches to explanation. He had to adopt the more moderate stance because to endorse the more extreme attitude seemingly advocated by his 1979 rhetoric would have placed him in an untenable position. Biological science cannot advance without the adaptationist program, and Gould knew it.

A representatively reasonable evaluation of the 1979 essay was written by Ernst Mayr nine years later. While endorsing Gould and Lewontin's cautions against the overuse and misuse of the program, Mayr reminded the leftists not to throw out the baby with the bathwater:

> Considering the evident dangers of applying the adaptationist program incorrectly, why are the Darwinians nevertheless so intent on applying it? The principal reason for this is its great heuristic value. The adaptationist question, "What is the function of a given structure or organ?" has been for centuries the basis for every advance in physiology. If it had not been for the adaptationist program, we probably would still not know the functions of thymus, spleen, pituitary, and pineal.[79]

The overall three-decade tone of Gould's writing is to endorse just this position. Then why was he so vehement in the 1979 essay? To try to explain why Gould's short-run rhetoric could be so at variance with his long-run position, I must engage in a little amateur psychoanalysis. Gould knew, in his heart of hearts and brain of brains, that biology could not jettison adaptationism. Yet, in the large political part of his character, he feared the consequences if the program were to be turned with full force on human behavior. Social practices that he execrated, he believed, would become "naturalized" so as to seem inevitable, and thereby ultimately defensible. In his effort to keep adaptationism functioning as part of biology while preventing it from being applied with reckless enthusiasm to humans, Gould followed a strategy that involved both rhetorical and intellectual inconsistency.

If the reasonable position is that adaptationism, as a methodological stance, is not only defensible but essential, why do sociobiologists frequently seem to fulfill Gould's worst fears? I confess that I do not know. The available examples seem convincing that sociobiologists often reach conclusions that reinforce status quo social relationships. But I cannot see why approaching human behavior from the standpoint of adaptationism must unavoidably lead to conclusions that whatever is, must be. Why cannot sociobiological assumptions lead to leftist conclusions? Is it only because leftists do not like sociobiology, and therefore refuse to adopt it as a method? What if they did?

On this point, a passage in Ruth Bleier's *Science and Gender* is instructive. Bleier is at least as far left as Gould—she believes that "the categories and meanings of 'women' and 'men' are cultural or symbolic constructions rather than reflections of biological 'givens.'"[80] She detests sociobiological research. At one point, she lampoons the genre's methodology with a creative reconstruction:

> If I were for the moment to accept Sociobiological premises, my predictions would be quite different from those proposed by Sociobiologists. Since women have a great biological investment in each pregnancy, which predisposes them to provide most of the parental care in order to protect optimally their genes in their offspring, I would predict:
>
> 1. A low incidence or absence of postpartum depression in women, since depression is *not* the optimal mental/physical state for the high energy requirements of postpartum lactation and infant care. . . .
> 2. A high incidence of postpartum depression in fathers, because they are deprived of a considerable portion of the parental care formerly invested in them by their wives. . . .
> 3. A low incidence of depression in women in general, since most of them are fulfilling their biological predispositions to mothers and nurturers. . . .[81]

Bleier intends these "predictions" to be taken as satire, but if instead we take them seriously we see that she raises good questions. Her predictions are eminently testable, and we already know the answers—nature falsifies her satirical tests. Women are more depression prone than men. The question then becomes why such logically derived predictions have been falsified.

Where is the flaw in the premises that leads to such empirical results? To ask such a question is to imply a research program. It seems to me that Bleier has a bright future as a sociobiologist, albeit of the left.

In other words, I see no inherent reason why leftist research could not proceed from sociobiological assumptions. Yet the work produced so far seems to be generally supportive of the point of view that various kinds of human inequalities are natural because they are adaptive. Gould's characterization of the political implications of the sociobiological method is unconvincing, yet much of the research arising from that method at least partly fulfills his expectations. I declare this contest a draw.

Two Morals

What have we learned from the fracas over the political implications of the biological controversy concerning human nature? First, no matter how much some scientists insist that science has political implications, other scientists will not only continue to deny the interpretation but be offended by it. Conversely, no matter how much some scientists insist that their motivations are only to learn the truth, and that they ruthlessly exclude political calculations from their analyses, other scientists will not only continue to accuse them of farcical naivety, but of bad faith.

This may be a dispiriting conclusion, because it leads to the expectation that scientific disputes will always take place within a context of rancorous misunderstanding and personal spite. But perhaps that attitude rests on an incomplete understanding of the nature of science. Philosopher David Hull has argued that "infighting and personal vendettas" are always a part of the scientific process, but that "factionalism, social cohesion, and professional interests need not frustrate the traditional goals of knowledge-acquisition."[82] Hull views the personal animosity that accompanies much scientific discussion as an analogue to the competition among economic producers that Adam Smith showed long ago could redound to the public interest. In Hull's view, the intense personal conflict that accompanies the scientific process might be bad for individual scientists, but it is good for the rest of us because it expands the fund of scientific knowledge. If Hull is right, then perhaps the permanent and irreconcilably incompatible ontological stances of different biological factions, while they will forever frustrate evolutionary biologists themselves, may stimulate greater total knowledge in the long run.

Perhaps. If the leftists are right, however, the fund of scientific knowledge that is ever expanding is always going in the wrong direction, and always reinforcing basic social injustices. Only if the scientists who insist that they can separate politics from science are right does Hull's formulation work. If Hull is wrong, then the permanent schism between the advocates of politicized science and the advocates of nonpoliticized science will simply be one more loud, disagreeable facet of the general social deterioration.

Second, the scientific conflict over the essence of human nature is a battle fought largely at the level of implication. Not a single scientist working in the field of sociobiology, or related fields such as evolutionary psychology, will admit to pursuing research in order to reinforce racist, sexist, or capitalist social structures. When these scientists proclaim their personal political goals, it is always to endorse the ideals of liberal democracy. Edward Wilson, the founder of the enterprise, asserted, "We are not compelled to believe in biological uniformity in order to affirm human freedom and dignity."[83] David Buss, whose work reinforces almost every known sexual stereotype, informed us, "If I have any political stance on issues related to the theory, it is the hope for equality among all persons regardless of sex . . . and regardless of preferred sexual strategy."[84] J. Philippe Rushton, whose work reinforced almost every known racial stereotype, wanted us to understand that "there are no necessary policies that flow from race research. . . . Ultimately, the study of racial differences may help us to appreciate more fully the nature of human diversity as well as the binding commonalities we share with other species."[85]

I accept the sincerity of Wilson, Buss, and Rushton, and all other sociobiologists, in abjuring any but the most benign political intentions. I think the leftists often make a mistake in accusing the sociobiologists of consciously trying to pursue an agenda that will provide ammunition to conservative forces in society. The personal polemics of the leftists tend not only to be unconvincing, but, much worse, to distract attention from the far more weighty charge that somehow the *implications* of the ontology and methodology of sociobiology lend themselves to the support of social elites. For the fact is that it is the conservative forces in society—the racists, sexists, and capitalists, although not, of course, the religious zealots—who bring sociobiological conclusions to the attention of the public, with evident approval.

Neo-Nazis in Britain have cited the writings of John Maynard Smith and Richard Dawkins to justify their racial nationalism.[86] Members of the extreme right in France have openly embraced Wilson's ideas.[87] In the United States, a conservative journalist writing in the *National Review* cited "the new biological learning" as supporting an economy based on self-interest, the traditional differences between the sexes, and "natural inequality."[88]

Sociobiologists cannot be blamed for the use, or misuse, of their ideas, any more than Gould can be blamed for the misuse of his ideas by Christian creationists. Nevertheless, the fact that it is always the right, never the left, that appeals to such research must draw our attention to the practical implications of scientific points of view. I am not convinced by Gould, Lewontin, and Bleier that there is a necessary *scientific* connection between an adaptationist methodology and conservative politics. But I think that the *political* connection between the two is unarguable. Whatever the intentions of scientists, reality seems to demand that scientific study into human nature will have political consequences. Given that fact, perhaps the attention given to leftist critiques of sociobiology is justified.

notes

1. George H. Sabine, *A History of Political Theory* (New York: Henry Holt, 1950), pp. 119–21.

2. Thomas Hobbes, *Leviathan: Or the Matter, Forme and Power of a Commonwealth Ecclesiastical and Civil* (New York: Collier Books, 1962), pp. 104, 119, 132, 135, 143, 167.

3. John Rawls, *A Theory of Justice* (Cambridge: Harvard University Press, 1971), pp, 19, 60, 101, 154–55.

4. Ruth Bleier, *Science and Gender: A Critique of Biology and Its Theories on Women* (Oxford: Pergamon Press, 1984), p. vii.

5. Ullica Segerstrale, *Defenders of the Truth: The Sociobiology Debate* (Oxford: Oxford University Press, 2000), pp. 2, 41–42, 77, 120, 206.

6. Edward O. Wilson, *On Human Nature* (Cambridge, MA: Harvard University Press, 2004), p. xvi; John Alcock, *The Triumph of Sociobiology* (Oxford: Oxford University Press, 2002), p. 221; Richard Dawkins, *Unweaving the Rainbow: Science, Delusion and the Appetite for Wonder* (Boston: Houghton Mifflin, 1998), p. 193–203; Daniel C. Dennett, *Darwin's Dangerous Idea: Evolution and the Meanings of Life* (New York: Simon and Schuster, 1995), pp. 262–67.

7. E. E. Schattschneider, *The Semi-Sovereign People: A Realist's View of Democracy in America* (New York: Holt, Rinehart, and Winston, 1960), p. 68.

8. Richard C. Lewontin, *Biology as Ideology: The Doctrine of DNA* (Concord, ON: House of Anansi, 1991), p. 97.

9. Richard Lewontin, in conversation with the author by phone, October 31, 2006.

10. Steven Pinker, *The Blank Slate: The Modern Denial of Human Nature* (New York: Viking, 2002), p. 73.

11. Wilson, *On Human Nature*, p. 167.

12. Stephen Jay Gould, *An Urchin in the Storm: Essays about Books and Ideas* (New York: W. W. Norton, 1987), pp. 114, 115.

13. Alcock, *Triumph of Sociobiology*, pp. 130, 159.

14. Ernst Mayr, *Toward a New Philosophy of Biology: Observations of an Evolutionist* (Cambridge, MA: Harvard University Press, 1988), pp. 129–30.

15. Ibid., p. 151.

16. Richard Dawkins, *The Extended Phenotype: The Long Reach of the Gene* (Oxford: Oxford University Press, 1999), p. 31.

17. Michael Ruse, *Darwin and Design* (Cambridge, MA: Harvard University Press, 2003), p. 218.

18. Stephen Jay Gould, *The Structure of Evolutionary Theory* (Cambridge, MA: Harvard University Press, 2002), p. 268.

19. Stephen Jay Gould and Richard C. Lewontin, "The Spandrels of San Marco and the Panglossian Paradigm: A Critique of the Adaptationist Paradigm," *Proceedings of the Royal Society of London* B 205 (1979): 594.

20. Stephen Jay Gould, *Ever Since Darwin: Reflections in Natural History* (New York: W. W. Norton, 1977), pp. 79–90.

21. Stephen Jay Gould, *Eight Little Piggies: Reflections in Natural History* (New York: W. W. Norton, 1993), pp. 27–29.

22. Edward O. Wilson, *Consilience: The Unity of Knowledge* (New York: Random House, 1999), pp. 74, 291.

23. Wilson, *On Human Nature*, p. 34.

24. Richard C. Lewontin, *The Triple Helix: Gene, Organism, and Environment* (Cambridge, MA: Harvard University Press, 2000), p. 48; see also Richard Levins and Richard Lewontin, *The Dialectical Biologist* (Cambridge: MA: Harvard University Press, 1985), pp. 65–84.

25. Gould, *Structure*, p. 1174.

26. Stephen Jay Gould, "Fulfilling the Spandrels of World and Mind," in Jack Selzer, ed., *Understanding Scientific Prose* (Madison: University of Wisconsin Press, 1993), p. 320.

27. Charles Bazerman, "Intertextual Self-Fashioning: Gould and Lewontin's

Representation of the Literature," in Jack Selzer, ed., *Understanding Scientific Prose* (Madison: University of Wisconsin Press, 1993), pp. 29, 32, 37.

28. Carolyn R. Miller and S. Michael Halloran, "Reading Darwin, Reading Nature; Or, On the Ethos of Historical Science," in Jack Selzer, ed., *Understanding Scientific Prose* (Madison: University of Wisconsin Press, 1993), p. 118.

29. Jeanne Fahnestock, "Tactics of Evaluation in Gould and Lewontin's 'The Spandrels of San Marco,'" in Jack Selzer, ed., *Understanding Scientific Prose* (Madison: University of Wisconsin Press, 1993), p. 161.

30. Ibid., pp. 168–69.

31. Gould and Lewontin, "Spandrels of San Marco," p. 581.

32. Ibid., p. 582.

33. Gould, *Structure*, p. 1259.

34. Ibid., p. 583.

35. Gould and Lewontin, "Spandrels of San Marco," pp. 584–85.

36. Ibid., p. 584.

37. Ibid., p. 585.

38. Ibid., p. 594.

39. David C. Queller, "The Spandrels of St. Marx and the Panglossian Paradox: A Critique of a Rhetorical Programme," *Quarterly Review of Biology* 70, no. 4 (December 1995): 485–89.

40. John Tooby and Leda Cosmides, "The Psychological Foundations of Culture," in Jerome H. Barkow, Leda Cosmides, and John Tooby, eds., *The Adapted Mind: Evolutionary Psychology and the Generation of Culture* (New York: Oxford University Press, 1992), p. 102.

41. Jeffrey S. Levinton, *Genetics, Paleontology, and Macroevolution*, 2nd ed. (Cambridge: Cambridge University Press, 2001), p. 227.

42. Dennett, *Darwin's Dangerous Idea*, pp. 238, 247.

43. Timothy Shanahan, *The Evolution of Darwinism: Selection, Adaptation, and Progress in Evolutionary Biology* (Cambridge: Cambridge University Press, 2004), p. 168.

44. Kim Sterelny, *Dawkins vs. Gould: Survival of the Fittest* (Cambridge: Icon Books, 2001), pp. 54–55.

45. Gould, *Urchin in the Storm*, p. 121.

46. Stephen Jay Gould, *Full House: The Spread of Excellence from Plato to Darwin* (New York: Three Rivers Press, 1996), p. 196.

47. Stephen Jay Gould and Elisabeth S. Vrba, "Exaptation—A Missing Term in the Science of Form," *Paleobiology* 8, no.1 (1982): 6.

48. Gould, *Structure*, p. 1254.

49. Edward O. Wilson, *Sociobiology: The New Synthesis*, 25th anniversary edition (Cambridge, MA: Harvard University Press, 2000).

50. Ibid., p. 4.
51. Ibid., p. 3.
52. Ibid., p. 5.
53. Ibid., p. 4.
54. Ibid., pp. 550, 575.
55. Edward O. Wilson, *Naturalist* (Washington, DC: Warner Books, 1994), pp. 331–32.
56. See for example Barkow, Cosmides, and Tooby, *Adapted Mind*; John R. Alford, Carolyn L. Funk, and John R. Hibbing, "Are Political Orientations Genetically Transmitted?" *American Political Science Review* 99, no. 2 (May 2005): 153–68.
57. E. Allen, B. Beckwith, J. Beckwith, S. Chorover, D. Culver, M. Duncan, S. J. Gould, R. Hubbard, H. Inouye, A. Leeds, R. Lewontin, R. Madansky, C. Miller, R. Pyeritz, M. Rosenthal, and H. Schreir, "Against 'sociobiology,'" *New York Review of Books* 22 (1975): 43–44.
58. Ibid.
59. Segerstrale, *Defenders*; see also Pinker, *Blank Slate*, pp. 105–35; Wilson, *Naturalist*, pp. 307–53; Gould, *Urchin in the Storm*, pp. 107–23; R. C. Lewontin, Steven Rose, and Leon J. Kamin, *Not in Our Genes: Biology, Ideology, and Human Nature* (New York: Pantheon Books, 1984), pp. 3–83, 233–64; Philip Kitcher, *Vaulting Ambition: Sociobiology and the Quest for Human Nature* (Cambridge, MA: MIT Press, 1985); John Alcock, *The Triumph of Sociobiology* (Oxford: Oxford University Press, 2001).
60. Stephen Jay Gould, *Hen's Teeth and Horse's Toes: Further Reflections in Natural History* (New York: W. W. Norton, 1983), pp. 242–43.
61. Wilson, *On Human Nature*, p. xvi.
62. Alcock, *Triumph*, p. 44.
63. Pinker, *Blank Slate*, p. 114.
64. Steven Weinberg, *Facing Up: Science and Its Cultural Adversaries* (Cambridge, MA: Harvard University Press, 2001), p. 118.
65. Wilson, *On Human Nature*, p. 143.
66. Albert Somit and Steven A. Peterson, *Darwinism, Dominance, and Democracy: The Biological Bases of Authoritarianism* (Westport, CT: Praeger, 1997), p. 3.
67. David Barash, *The Whisperings Within: Evolution and the Origin of Human Nature* (New York: Penguin, 1979), pp. 189–90.
68. J. Phillipe Rushton, *Race, Evolution, and Behavior: A Life History Perspective* (New Brunswick, NJ: Transaction, 1995), p. 255.
69. Gad Saad, "Evolution and Political Marketing," in Albert Somit and Steven Peterson, *Human Nature and Public Policy: An Evolutionary Approach* (New York: Macmillan, 2003), p. 121.

70. David M. Buss, *The Evolution of Desire: Strategies of Human Mating* (New York: HarperCollins, 1994), p. 8.

71. Ibid., p. 3.

72. Ibid., p. 19.

73. Ibid., p. 4.

74. Ibid., p. 62.

75. Ibid., p. 82.

76. Ibid., p.154.

77. Stephen Jay Gould, *The Mismeasure of Man* (New York: W. W. Norton, 1981), p. 28.

78. Kitcher, *Vaulting Ambition.*

79. Mayr, *Philosophy of Biology*, p. 153.

80. Bleier, *Science and Gender*, p. 72.

81. Ibid., pp. 39–40.

82. David L. Hull, *Science as a Process: An Evolutionary Account of the Social and Conceptual Development of Science* (Chicago: University of Chicago Press, 1988), p. 26.

83. Wilson, *On Human Nature*, p. 50.

84. Buss, *Evolution of Desire*, p. 18.

85. Rushton, *Race, Evolution, and Behavior*, p. 257.

86. Hull, *Science as a Process*, p. 230.

87. Bleier, *Science and Gender*, p. 9.

88. Quoted in Stephen Jay Gould, *I Have Landed: The End of a Beginning in Natural History* (New York: Three Rivers Press, 2003), p. 220.

Chapter 5

science and human inequality

When Thomas Jefferson wrote "All men are created equal" in the Declaration of Independence in 1776, he meant moral, not material equality. On his mind was the *Second Treatise of Government* by English philosopher John Locke from 1690, the argument of which dominated the thinking of eighteenth-century Americans. Locke had been completely clear that the equality of all men was given by God, making it impossible to abrogate, but that this divine gift did not extend to equality of property.[1] Jefferson was not contradicting his earlier self, then, when he opined in a letter to John Adams in 1813 that "there is a natural aristocracy among men. The grounds of this are virtue and talents."[2] He saw no inconsistency between, on the one hand, giving all men the right to vote and equality before the law, and on the other, allowing the more talented to rise to riches and power while the less talented either stayed in their low position, or even sank.

Jefferson was a leftist for his society and his era, that era being the last quarter of the eighteenth century and first quarter of the nineteenth. Then and for some time afterward, the American left contented itself with creating equality of opportunity by eliminating both hereditary and government-created positions of power and wealth. For much of American history, it did not seem to occur to most leftists that if equality of opportunity were ever to be achieved, the different varieties of human competence, plus luck, would continue to create inequalities of result.

European leftists picked up on the possibilities more quickly than Americans. When, in 1840, the Frenchman Pierre Proudhon asked rhetorically, "What is property?" and rhetorically answered, "Property is theft!" he began a tradition of thought that had made socialism a force in European politics by the turn of the twentieth century.[3] Socialism, with its attendant beliefs and values, did not become a major player on the American intellectual scene until the 1930s, but once established, it became important among scholars and the people in society who pay attention to scholars. By the 1970s the American left, while not dominated by socialism, had become heavily influenced by certain strands of socialist thought.

Chief among the socialist values that had become very important in the thinking of leftists was *equality of result*. The "virtue and talents" that Jefferson had believed justified unequal distribution of property were increasingly seen by leftists as the consequence of chance, and therefore undeserved. When, as I have chronicled in chapter 3, Gould argued that all of evolution was governed by contingency, he raised the leftist emphasis on the lottery-like essence of the distribution of human talent to a principle of nature.

Given their belief in the unjustified basis of all inequalities, leftists emphasized the moral imperative of redistribution of wealth and power in all circumstances and all societies. Leftist biologists naturally looked to their own profession to increase public understanding of the threats to equality that they believed were inherent in present social practices. The statement by the Marxists Richard Lewontin, Steven Rose, and Leon Kamin from 1984 is more extreme than a position that would be endorsed by most American liberals, but it is a fair summary of the position argued by Gould in many of his publications:

> The ideology of equality has become transformed into a weapon in support of, rather than against, a society of inequality by relocating the cause of inequality from the structure of society to the nature of individuals. First, it is asserted that the inequalities in society are a direct and ineluctable consequence of the differences in intrinsic merit and ability among individuals. . . . Second . . . biological determinism locates . . . successes and failures . . . as coded, in large part, in an individual's genes. . . . Finally, it is claimed that the presence of such biological differences between individuals of necessity leads to the creation of hierarchical societies.[4]

Nevertheless, a large proportion of the American public continued to believe in the principle of reward according to merit. And the merit, or talent, that was perhaps most revered through all levels of society was intelligence. Highly intelligent people, in late twentieth- and early twenty-first-century America, were seen as the closest equivalent to Jefferson's natural aristocracy. And whether or not intelligence establishes a natural elite in some ultimate sense, there is no doubt that in advanced industrial society in general, and American society in particular, high intelligence gives a person access to elite rewards. Intelligence test scores are the best predictors of economic success in Western society.[5]

As in much else that Gould wrote, he was not completely consistent when addressing the several types of issues associated with the idea of equality. While frequently issuing such broad rhetorical statements as "human equality is a contingent fact of history,"[6] he was quite capable of making fine distinctions on the subject. He believed, for example, that "equality is a magnificent system for human rights and morality in general, but not for the evaluation of evidence,"[7] and, more importantly:

> We too often, and tragically, confuse our legitimate dislike of elitism as imposed limitations with an argument for leveling all concentrated excellence to some least common denominator of maximal accessibility. . . . Elitism is repulsive when based upon external and artificial limitations like race, gender, or social class. . . . But if only a small minority respond, and these are our best and brightest of all races, classes, and genders, shall we deny them the pinnacle of their soul's striving because all their colleagues prefer passivity and flashing lights? . . . What is wrong with this truly democratic form of elitism?[8]

At the level of rhetoric, Gould was therefore much like Jefferson, endorsing equality in a general sort of way but also embracing an elitism of the talents. Nevertheless, the test of any philosophy is in its practical application, and the particular practical application that attracted Gould's attention was intelligence testing.

Measuring Talent

Mental testing to identify bright students and weed out the intellectually challenged was first instituted in America and Western Europe in the late nineteenth century, and it began to be subject to statistical controls in the first decade of the twentieth. From the beginning of the effort of scientists to bring quantitative precision to intelligence testing, the methodology was the subject of intense wrangling. During the 1920s and 1930s, one of the eminent testers of the day, Charles Spearman, engaged in a scholarly debate, conducted in the professional journals, with two other founders of the discipline, Edward

Thorndike and Louis Thurstone. The disagreement was over whether "intelligence" was best defined as a single entity or multiple entities.

Spearman believed that the methodology of factor analysis, which he invented, showed that there was, essentially, one general intellectual ability, which he termed *g*. As with general athletic ability, this general intellectual ability could be expressed in different specific abilities. As one good athlete might have a special aptitude for running and swimming, and another for throwing and weight lifting, different people with roughly comparable general intelligence might nevertheless be cognitively stronger at some mental tasks than at others. Spearman argued that any particular test measured two factors, *g* and a specific factor unique to that test. He suggested that the best way to measure any given individual's intelligence was to administer a battery of tests to that person. The common factor in all the tests would be *g*, and the effects of the specific factors would tend to cancel each other out.[9]

In opposition, Thorndike and Thurstone argued that the correct use of the same methodology pointed to multiple factors, and that therefore there was no such thing as *g*—there was no general intelligence, only separate cognitive abilities. In 1920 Thorndike decided that there were eight important realms of mental ability, summarized as spatial visualization, perceptual ability, verbal comprehension, numerical ability, memory, word fluency, and both inductive and deductive reasoning capacity.[10] During the 1930s, Thurstone came up with a very similar list.[11] But the controversy over whether to describe intelligence in the singular or plural has never been resolved to the satisfaction of all psychologists.

As intelligence testing developed over the course of the twentieth century and into the twenty-first, the split in scientific circles between those who saw their subject as one general mental ability with a variety of specific manifestations, and those who saw it as a cornucopia of diverse and independent talents, not only continued but expanded. In 1986, when Robert Sternberg and Douglas Detterman surveyed two dozen psychologists and psychometricians (specialists in mental measurement) about how they conceptualized "intelligence," the two researchers collected two dozen definitions.[12] Various psychologists have come up with three, or a dozen, or twenty cognitive abilities that, they argue, should be considered as different types of intelligence. The record seems to be held by J. P. Guilford, who has postulated some 120 intelligence factors.[13]

Moreover, scholars in the field keep modifying the understanding of intelligence in sophisticated new ways, which naturally expands the number of concepts the specialists must keep track of. Researchers now speak of an "emotional intelligence" that is related to, but not the same as, intellectual intelligence.[14] They work with the idea of "social intelligence," which seems to have three domains—social understanding, memory, and knowledge.[15] They have decided that cognitive intelligence consists of two types, "fluid intelligence," which reflects higher mental abilities, especially reasoning, and "crystallized intelligence," which reflects the knowledge acquired from life experiences.[16] It would not be too much of an exaggeration to summarize the current scientific understanding of intelligence as consisting of as many conceptions as there are scientists studying the phenomenon.

Yet for all its heterogeneous splendor, the study of intelligence does display some common ideas and central tendencies. When Snyderman and Rothman surveyed those working in the general field during the mid-1980s, they reported that 58 percent of their respondents favored some form of definition that rested on an understanding of the concept as a single, general component.[17] If the 13 percent of nonresponders are omitted, then a full two-thirds endorsed the notion that *g*, or some similar concept, existed. Meanwhile, of the total sample, 13 percent responded that intelligence consisted of multiple faculties, while 16 percent held that the evidence was so ambiguous that no conclusion could be reached.[18] Among professionals, then, a clear majority endorsed the concept that intelligence is a single entity, but the dissenting minority was so large that it should convince all nonprofessionals to avoid dogmatism when discussing this subject.

The majority consensus was based on the fact that the several types of intelligence tests all tended to produce scores that were broadly predictive of success in school and college, in armed forces training programs, in job performance in the private sector, and even the likelihood of turning out to vote on election day.[19] These general findings have been replicated and reported in scientific, peer-reviewed articles about research in Europe, Africa, South America, and Asia.[20] Additionally, *g* as measured on intelligence tests tends to correlate consistently with moderate strength (about .40) with nonverbal mental measures such as tests of reaction time to stimuli.[21] "Intelligence," then, as measured by modern tests, does not seem to be a phantom but a real phenomenon, however difficult it is to define, and however many different ways scholars have invented for studying it.

If the actual existence of some sort of general intelligence is accepted, then the next issue is the one of determining what causes it. There are two candidates. Either mental abilities are the result of genes or of the environment—or of some interaction between the two. In practice, almost all scholars assume that the existence of intelligence is the result of an interaction; the controversy is over the relative strength of nature versus nurture. And here the very idea of measurable intelligence becomes acutely political.

For if, on the one hand, intelligence is largely the work of genes, and if it can be shown that higher intelligence leads to greater wealth and status, such a factual state will encourage political conservatives to argue that the people on the top of society are just where they ought to be. Similarly, if intelligence is genetically determined, and people of generally lower intelligence tend to cluster at the bottom of the occupation and status scale, then conservatives will be emboldened to argue that those people are also just where they should be. Gould and other leftists believe that such a situation would encourage the people who want to "make nature herself an accomplice in the crime of political inequality," in Condorcet's words.[22]

On the other hand, if the hereditarian contribution to intelligence is small, and the environmental contribution correspondingly large, then the left thinks that two conclusions would be warranted. First, people's performance can be raised. Presumably, the right public policies, especially the right investment of social resources, could greatly improve the intelligence of those who are now least intelligent. Second, those on the bottom of the social hierarchy are, in essence, victims of past public policy, a situation that would seem to justify redistributive policies by government to right the balance.

In the United States, this argument over the relative contribution of genes and environment to intelligence is often combined with the national argument over race, but it is logically separate. Black citizens as a group tend to score about a standard deviation lower than white citizens on all intelligence tests, so that, when Gould was writing about this subject, the average IQ of blacks was roughly 85, and of whites, roughly 102.[23] The frequently ferocious American controversy over various public policies relevant to race differences is perfectly congruent with the argument over the genetic component of intelligence. Conservatives, who are skeptical of government programs to attempt to improve the measured intelligence of African Americans, emphasize the proportion of the concept caused by genes. Liberals, progressives, or socialists, who support greater government effort to raise black measured intelligence, emphasize the proportion of the concept caused by environmental factors. Because of America's violent racial history and touchy present racial political situation, the controversy over the differing contributions of nature and nurture to variance in intelligence usually gets around to race sooner or later. Nevertheless, the argument is independent of race and would continue to divide the left from the right even if the race conflict disappeared.

The political argument is made even more productive of resentful polemics because of the frequent misunderstanding over the difference between group averages and individual capacity. When psychometricians discuss the results of tests, they speak of some central tendency of the scores of groups. The groups can be defined in any number of ways: by race, or age, or sex, or occupation, or grade in school, or by any definition the tester finds interesting. The averages reported are just that, summaries of aggregate tendencies of many individual scores. The differences that are discussed—the amount of genetic versus environmental causation, for example—apply to the groups, not to the individuals within the groups. A psychologist who argues that intelligence is "half inherited and half from the environment" does not believe that any one person inherited half her intelligence and acquired half of it from her family upbringing, schooling, and peer groups.

Moreover—and here is where the political misunderstanding becomes explosive—a finding that the members of one group average a higher IQ than the members of another does not tell us anything about the individual members of either group. The fact that whites as a group have a higher average IQ than blacks as a group does not mean that any given white person is necessarily more intelligent than any given black person. Such questions can only be answered by consulting the score of each individual. Nevertheless, political rhetoric being what it is, lower and higher group averages are bound to be employed by various right-wing interests to justify invidious distinctions between the individuals of each group. That being so, left-wing ideologues are bound to react with reflexive negativity to any discussion of group averages.

The study of intelligence is thus deeply political, even if the scholars who specialize in the field have the purest scientific motives. Sometimes psychologists write articles justifying their interest in the topic, and seeking to separate the scientific aspects of their study from the political aspects. Thus, John Loehlin has written:

> The study of race differences in intelligence, especially if done from a biological perspective, requires some courage. . . . But the issues—many of them—remain important. And the questions, at least some of them, should be answerable. So they ought to be asked.[24]

Any scientist can sympathize with Loehlin's desire to be left alone in the sincere and honest pursuit of truth. But as long as scientific findings, however tentative and provisional they may be, have implications for the relationship of millions of people in a democratic society, the outside world will be a looming presence in the institution of science.

Disagreements among scientists, then, are directly relevant to disagreements among people with less pure motives. A number of psychologists argue that the contribution of genetic differences to variations in *g* is not only large but basically unalterable. That is, implicitly or explicitly they maintain that the amount of variance the environment contributes to measured intelligence is so small that efforts to raise IQ by enriching environmental conditions are largely futile. The most famous of the psychologists who argue the hegemony of genetic intelligence is Arthur Jensen, whose 1969 article "How Much Can We Boost IQ and Scholastic Achievement?" set off a mini-explosion in the world of political wonkery.[25] But Jensen is not the only scholar who argues that the weight of the evidence must compel the conclusion that a good deal more than half of all the variation in cognitive ability within a given population (or at least within a given Western population) must be the result of variations in genetics.[26]

There is enough counter-evidence, however, to persuade sensible observers who are not experts in genetics to pause before endorsing any conclusion. On one front, for example, in 2005 Duckworth and Seligman reported a study in which they determined that self-discipline, as measured early in the school year, accounted for more than twice as much variance in students' grades as did IQ.[27] This study tends to diminish the importance given to *g* as a predictor of behavior. On another front, in 2004 Schellenberg published his results in causing small increases in IQ by giving children music lessons.[28] Such findings suggest that rather mild interventions in the environment can have a positive effect on intelligence. Similarly, in 2006 Dickens and Flynn reported that African Americans gained, on the average, 4 to 7 IQ points on non-Hispanic whites between 1972 and 2002.[29] Presumably, the gains are the result of improvements in the social conditions of black citizens, or, in other words, in the environment. Such findings are not definitive, and will undoubtedly be modified with further research, but they tend to instruct prudent nonexperts to treat all expert findings with respectful skepticism.

Every expert, including Gould,[30] believes that intelligence is determined by an interaction between heredity and environment. The serious research that has been done on the topic attempts to discover the contributions of each cause to variations of intelligence within populations.

On the one hand, psychologists have attempted to follow good scientific procedure by holding socioeconomic status (SES) constant while studying the contribution of genes, but these variables are fairly coarse and certainly do not cover every influence of the environment. One of the findings of IQ research, for example, is that nutrition while the fetus is in the womb, and for the first six months in a baby's life, are important determinants of later intelligence scores.[31] Since SES variables usually do not control for early malnutrition, they cannot protect the results of such studies from overemphasizing the hereditarian component of IQ. There are undoubtedly many more unmeasured variables outside of SES controls, which is why such research usually generates rather low predictability.

On the other hand, it is fairly easy to hold inheritance constant while varying environment. Researchers test identical twins (sharing exactly the same set of genes) who have been raised apart or together, and fraternal twins (sharing about half their genes) who have been raised apart or together. The scholars then put the test scores through a series of statistical procedures to come up with measures of the extent to which the environment contributes to variations in the twins' IQs.[32] The results are useful, but they still depend upon being able to identify and measure the magnitude of the environmental inputs. Researchers try to measure environmental differences by taking account of family education, income, number of books in the home, and so forth. But inevitably, many variables go unmeasured. As Lewontin, Rose, and Kamin put the same point:

> The most striking and consistent observation in adoption studies is the raising of IQ, irrespective of any correlation with adoptive or biological parents. The point is that adoptive parents are not a random sample of households but tend to be older, richer, and more anxious to have children. . . . So the children they adopt receive the benefits of greater wealth, stability, and attention.[33]

Psychologists who view *g* as a stable, measurable entity, and who conduct studies on its heritability (generally on identical twins reared apart or together), usually report a genetic component of between 50 and 80 percent.[34] Most of these studies have been conducted in the United States or Britain, where presumably environments are relatively similar, and benign. A few have been conducted in various third-world countries, but these tend to lack the sorts of controls social scientists would prefer. Black respondents from sub-Saharan Africa, for example, tend to score quite low on intelligence tests—with an IQ of about 70.[35] One can only speculate on the sorts of environmental variables—including prenatal malnutrition—that are not accounted for in these studies.

The suspicion that genetic reasoning based on twin studies is contaminated by unmeasured environmental variables is a feature not only of the literature on intelligence. In 2005, Alford, Funk, and Hibbing published an article in the *American Political Science Review* in which they argued that political ideology is largely an inherited—that is, genetic—feature of human cognition. Their argument was based on twin studies, employing the same methodology as that used for IQ research.[36] In 2008 Evan Charney published a critique of their methodology in *Perspectives on Politics*. He rejected Alford, Funk, and Hibbing's conclusions for a number of reasons, one of the main ones being that there is the "possibility or probability of significant environmental confounding effects" in twin studies.[37] Charney quoted a variety of geneticists expressing their doubts as to whether environmental forces can be controlled in such studies.[38] In reply, Alford, Funk, and Hibbing, and in a separate article Hannagan and Hatemi, defended their methodology and cited various studies in the psychological and genetic literature that they argued rendered Charney's objections unconvincing.[39]

In other words, the dispute over the methodological ambiguities of twin tests has been transferred whole from psychology into political science. And as in psychology, it has not been resolved. Moreover, the nasty tone of the intelligence-testing dispute has also made the leap, the pro-twin testers accusing Charney, in essence, of being antiscience for reasons of personal insecurity, and Charney responding with appropriate indignation. Political scientists can look forward to years of a rancorous squabble that duplicates the experience of psychology.

My purpose in summarizing some of the problematics of intelligence research, and now, ideology research, is not to relegate all such studies to the trash bin. Almost all social research is fraught with difficulties caused by bad data, missing data, uncontrolled contaminating variables, spurious correlations, and many similar plagues, but we don't stop trying to study interesting problems because of them. We forge on. We do not conclude, as do Lewontin, Rose, and Kamin, that IQ tests are rendered completely useless by such methodological problems. The vast accumulation of data from such tests over more than a century of work amounts to considerably more than nothing. And, as Philip Vernon argues in his defense of intelligence testing, "While we must admit the force of these arguments, they are really objections to almost any kind of psychological research."[40] True, but not all kinds of psychological research create the political problems that are part of intelligence research. The problems arise when citizens, politicians, and sometimes scholars themselves use the ambiguous and arguable results of intelligence tests as a basis for advocacy of this or that public policy. When ethereal and debatable tentatives are turned into polemical certainties, the concept of intelligence loses what little substance it has in science. Psychologists have forged on in studying intelligence, and good for them. But the rest of us should treat their conclusions with a dose of caution, and discourage others from using them to buttress political positions.

This brings me to Gould. In 1981 appeared *The Mismeasure of Man*, his contribution to the national conversation over IQ, genes, and race. In this book, as in so much of his other work, his rhetoric often outran his more considered positions on biopolitical issues. A careless reading of Gould's comparison of intelligence testing to the "biological determinism" of sociobiology, and his identification of such tests as creating "a limit posed from without, but falsely identified as lying within," could easily lead to the conclusion that he opposed every test of everybody forever.[41] But his more considered opinion was, as usual, more conflicted and complicated.

Gould used two major strategies in criticizing mental testing. First, he deliberately stopped his discussion of the various ways people had tried to come up with a procedure for measuring a human external in order to understand the internal capacity in the 1920s, without addressing modern testing. "I have said little about the current resurgence of biological determinism because . . . I thought that it would be more valuable and interesting to examine the original sources of the arguments that still surround us."[42] Thus Gould analyzed and ridiculed the theories of *polygenism* (the belief that the different races were separate species), *craniology* (the doctrine that brain size determined intelligence), *recapitulation* (the conviction that the adults of inferior human groups must be like the children of superior groups), *stigmata* (the notion that potential criminals could be recognized, before they had committed any crimes, with regard to certain telltale external signs—unusually large jaws, for example), and early intelligence testing.

In regard to intelligence testing, he drew a portrait of early twentieth-century white researchers who were panic stricken at the thought that inferior types of humans, such as blacks and Eastern European immigrants, would outbreed traditionally dominant American groups, thus diluting the gene pool and causing national decline.[43] Gould did not have to state the moral baldly, for it was implicit in every line: for more than a century and a half, every attempt to measure human mental capacity has been motivated by the desire to create invidious distinctions, fraudulently believed to be natural, that buttressed the structure of social dominance. Any reader who could take a hint would be primed to doubt the integrity and scientific credentials of all researchers who attempted to defend modern intelligence tests.

Gould's second strategy was just as important to his implied argument, and more interesting from a scholarly point of view. In keeping with his general position that science with obnoxious political implications should be combated primarily at the level of ontology and methodology, he addressed the assumptions underlying the main statistical approach to the study of intelligence, factor analysis.

When a variety of different tests are administered to groups of people, correct answers on certain subjects may hang together—be correlated. People who tend to get verbal questions right may also get mathematical questions right, for example. Or, people who correctly answer verbal questions may do poorly on tests of spatial manipulation, in which case their answers on the two types of tests would not be correlated. Factor analysis is a method of determining which variables are correlated and to what extent. If a variety of answers hang together in a correlation matrix, the researcher may theorize that these answers represent a single underlying dimension—a factor—that associates those answers conceptually. Or, the researcher may theorize that there are a number of underlying dimensions, and look for several factors in the answers. In dealing with factor analysis of any set of answers, researchers may "rotate," or reconfigure, the axes on which the correlations are plotted, to produce any number of factors, depending on the theoretical expectations of the scholar. In any case, the manner of construction of a factor analysis, and the number of types of variables selected for inclusion, largely determine the conclusions to be gleaned from it.[44]

As Gould discussed at some length, in the early history of intelligence testing there was a controversy between Charles Spearman, who insisted that there was one evident cognitive factor measured by tests and illustrated by factor analysis, and Louis Thurstone, who maintained that a properly rotated analysis of intelligence tests revealed several factors.[45] Gould's point was not that one or the other was correct, but that it doesn't matter—because whether intelligence is measured as one entity or several, it is still just an artifact of the method used to measure it. Psychologists who regarded "intelligence" as a thing in the mind, he argued, were committing the error of reification; they invented an abstract concept and deceived themselves into investing it with concrete reality. Intelligence, as measured by psychologists, is an imaginary entity imposed on sets of numbers by psychologists. The numbers do not actually measure anything in the mind; they are an artifact of the mathematical technique. As Gould put it:

> Perhaps the factor axes are not fixed in number, but subject to unlimited increase as new tests are added. Perhaps they are truly test-dependent, and not real underlying entities at all. The very fact that estimates for the number of primary abilities have ranged from Thurstone's 7 or so to Guilford's 120 or more indicates that the vectors of mind may be figments of mind.[46]

In this passage is the heart, not only of Gould's criticism of intelligence testing, but of the scholarly left's in general. It contains what might be termed the political meta-argument of intellectuals who object to the potential social consequences of mental measurement. Gould maintains that the entire enterprise of such testing is based on an imaginary concept, a concept that is not found in nature but is the product of intellectual alchemy. As Levins and Lewontin worded the same point, IQ is "only a tautological relationship among a set of numbers" that psychologists delude themselves into thinking is a "revelation about the real world."[47]

There is some truth to this charge, but not as much as Gould, Levins, and Lewontin would want us to believe. It is true that the correlation matrices of factor analysis can be rotated in various ways, thereby allowing the researcher to interpret mathematical data according to their individual purposes. As the authors of one history of intelligence testing summarize, "The way that a factor analysis is conducted and the number and type of variables selected for inclusion largely determine what the analysis reveals."[48] But to admit that fact is merely to agree that data can always be looked at from different points of view, and that different analysts can sometimes interpret the same evidence in different ways. It is not to agree that the results of a factor analysis, any more than the results of any other form of statistical measurement, have no connection to empirical reality.

When trying to decide whether the charge of reification is true, the best question to ask is, "Do the results of factor analyses predict, with a fair amount of validity and reliability, human behavior in the real world?" If the answer is no, then the technique is, as Gould suggests, a figment of psychologists' minds. If the answer is yes, however, then the technique is measuring something actually there in the heads of real people. And, with the usual scientific cautions and acknowledgement of imprecisions, my own conclusion is that in general the answer is yes. Intelligence tests do predict, with tolerable validity and reliability, various kinds of human talents outside the tests themselves.[49] Therefore, the charge that "intelligence" is merely a reification with no substantive utility is mistaken. To say that intelligence actually exists, of course, is not to say that it is of great value, or that it is more important than other considerations, or that it is necessarily the most relevant topic to fight about, or that psychologists always correctly interpret the evidence of their tests, or that the mass media accurately report the findings of scientists, or that politicians fair-mindedly draw policy conclusions from media reports. But it is to decide that the left's charge that "intelligence" is a wholly imaginary construct is simply wrong.

The Mismeasure of Man was generally well reviewed when it appeared, and it continues to be cited in some books as a definitive refutation of much of the literature on intelligence.[50] But not surprisingly, many psychologists were unhappy with it, in 1981 and afterward. "It is a masterpiece of propaganda," wrote Steve Blinkhorn in a review in *Nature* in 1982. Gould had attacked "antique methods of factor analysis," giving no hint that "multivariate methods have progressed beyond Thurstonian techniques . . . no account of modern behavioral genetics . . . no hint of the current interest in cybernetic models."[51] When the book was reissued fifteen years later, J. Philippe Rushton accused Gould of "continuing a political polemic" and engaging in "character assassination of long deceased scientists whose work he misrepresents despite published refutations, while studiously withholding from his readers 15 years of new research that contradicts every major scientific argument he puts forth."[52] In 1999 Colin Cooper asserted that Gould's book "is widely regarded as a seriously misleading and partial book that is frequently factually incorrect."[53]

Gould had not written his book, of course, about modern psychological research on intelligence. Under ordinary circumstances it would be bad form in a review to criticize an author for neglecting to cover scientific activity that occurred long after the era he discussed in his book. But Gould's political approach had both invited the attack and rendered it appropriate. His strategy of implying conclusions about contemporary events without actually addressing them had made his neglect of modern thinking fair game.

Nevertheless, the outrage of some psychologists was irrelevant, because they were not Gould's intended audience. While partisans of *g* published their infuriated rebuttals to *The Mismeasure of Man* in scientific journals, Gould continued writing essays bashing intelligence testing in the *New York Review of Books*, *Natural History*, and his book-length collections of essays.[54] He contributed to the general uneasiness in educated circles with the very concept of intelligence testing. The general attack was so successful that when Colin Cooper wrote his textbook on the subject in the late 1990s, he tried to avoid using the very word *intelligence*, so questionable was its reputation.[55]

In this there were two ironies. First, the people who were most likely to have been persuaded by Gould and his allies to be suspicious of all references to IQ and intelligence tests were themselves high-IQ performers who had achieved their professional standing partly through their history of scoring impressively on such tests. The second irony is that Gould insisted that everyone who had read the book and interpreted it as blowing the whistle on all intelligence tests had misunderstood him. In the introduction to its second edition in 1996, he tried to set the record straight:

> My book . . . has been commonly portrayed . . . as a general attack upon mental testing. *The Mismeasure of Man* is no such thing, and I have an agnostic attitude (born largely of ignorance) toward mental testing in general. . . . *The Mismeasure of Man* is a critique of a *specific* theory of intelligence often supported by a *particular* interpretation of a *certain* style of mental testing: the theory of unitary, genetically based, unchangeable intelligence.[56]

Presumably, mental tests that measured multiple factors, environmentally caused, and changeable, would find favor with Gould. To repeat Richard Lewontin's mantra from the previous chapter, "The world must be malleable; human nature cannot be fixed. A reformist person must believe that changing people is possible."[57]

The Mismeasure of Man illustrated the way a reformist person could employ that most scientific of strategies, methodological critique, in pursuit of a political goal.

Thus, in regard to his putative profession, evolutionary biology, Gould's most important contribution to the debate over intelligence was to argue forcefully that evolutionary theory was *not* important to the study of that subject. Once scientists had realized, he urged, that the legacy of billions of years of natural selection on life on earth was to render human intelligence a function of environmental forces, and therefore malleable, then they should stop speculating about the adaptationist value of intelligence. Gould, the most famous evolutionist in America, was making an antievolutionist scientific argument that was, to use Ullica Segarstrale's term, "coupled" to his political argument.[58]

Second Verse

The controversy simmered quietly along for more than a decade after *The Mismeasure of Man*, only to boil over into a veritable volcanic explosion in 1994 with the appearance of Richard Herrnstein and Charles Murray's *The Bell Curve*.[59] Their book is an analysis of the results of the National Longitudinal Survey of Labor Market Experience of Youth, begun in 1979 with the testing of 12,686 Americans, aged 14 to 22.[60] The original respondents were reinterviewed periodically, so that the data Herrnstein and Murray analyzed contained information on childhood environments, parental socioeconomic status, educational achievements, work history, family formation, self-reported crimes, and several cognitive tests.

Over more than six hundred pages of analysis, Herrnstein and Murray reported that IQ scores predicted, better than any other measure, success at training for military specialties, success in private-sector jobs, tendency to live in poverty, school achievement, probability of being unemployed, probability of being divorced, likelihood of collecting welfare, probability of being convicted of a crime, and sundry related social behaviors.[61] Moreover, they argued that IQ was about 60 percent heritable, which, of course, left some but not a lot of room for changing people's lives by changing their environment.[62] Further, although much of their analysis was conducted on the white population only, they confirmed the general one standard deviation difference in the IQs of whites and African Americans, and acknowledged that "the social consequences are potentially huge."[63] "The evidence presented here should give everyone who writes and talks about ethnic inequalities reason to avoid flamboyant rhetoric about ethnic oppression."[64] In other words, the differences in income and social standing between blacks and whites in the United States were caused not by past or present racial discrimination but by the fact that blacks were, as a group, not as intelligent as whites.

Although Herrnstein and Murray were careful to state, "We cannot think of a legitimate argument why any encounter between individual whites and blacks need be affected by the knowledge that an aggregate ethnic difference in measured intelligence is genetic instead of environmental," they did not shrink from drawing a moral for public policy.[65] "It is time to return to the original intentions of affirmative action," to spend social resources to identify and nurture the careers of intelligent minority individuals, but to jettison the present workings of the policy, which consists of discriminating against whites and Asians while discriminating in favor of African Americans and Latinos.[66]

The Bell Curve not only became a bestseller and spawned a massive amount of public comment and political rhetoric, it also immediately stimulated hostile and semihostile reviews and critiques by scholars. Among those was an essay by Gould, who, drawing on his lifelong affection for the game of baseball, and referencing a common American folk saying that refers to an action or statement that is sleazily deceptive, gave his analysis the clever title, "Curveball."[67]

Herrnstein and Murray's book was tailor made to arouse Gould's ire, for it treated intelligence as a single, heritable entity that was relatively immune to change. His language in the review, as it had with *The Mismeasure of Man*, left no doubt as to his contempt for this sort of IQ research, labeling it "anachronistic social Darwinism" and "a manifesto of conservative ideology."[68] But also as with his earlier book, the bulk of his criticism was methodological.

In each of their chapters arguing a presumably causal relationship between measured IQ and some social behavior, Herrnstein and Murray had presented a statistical relationship known as a *regression curve*. For the relationship between intelligence and the likelihood of welfare dependency among white women, for example, they had drawn a chart showing any close reader that, for women with very low IQs the probability of going on public assistance within a year after the birth of their first child was almost 50 percent, whereas for women with very high IQs, the probability was well below 10 percent.[69] Similarly, the probability of committing a crime among white men dropped from almost 15 percent to virtually zero as one ascended the IQ scale, the probability of becoming unemployed similarly plunged, and so on.[70] All in all, it was an impressive graphic display that seemed to prove that intelligence played a very large part in determining the social behavior of everyone.

But after he had finished his obligatory ideological attack on Herrnstein and Murray, Gould proceeded with the far more significant job of showing that their science was unpersuasive. To repeat an observation I made in chapter 3, it is possible to fit a line to any set of data points, and there are a variety of statistical techniques for line fitting, depending on the purposes of the researcher. In the case of the regression methodology used by Herrnstein and Murray, the slope of the line is important, for it tells us how much of the movement in one variable (probability of being arrested) is caused by another variable (measured IQ). But the goodness of fit of the data points around the line, captured by the "r-squared" statistic, is equally important. It tells us how much of the total variation in the movement of the one variable is actually explained by the other variable. If the r-squared statistic is, say, .50, then half the variation in the probability of being arrested is explained by intelligence, and we have a relationship worthy of a book chapter. If the r-squared statistic is, say, .01, then only 1 percent of the probability of being arrested is explained by intelligence, and we would be less than honest to claim that IQ tells us anything important about being prone to crime.

As Gould pointed out with some force, Herrnstein and Murray failed to report the r-squared statistic for any of their regression curves in their main textual discussion, tucking them away in an appendix to their book. This was, he asserted, a "violation of all statistical norms I've ever learned."[71] This may seem to be a rhetorical overstatement on Gould's part, but in fact it is an accurate report on the situation in academic research. As Marvin Dunnette wrote in the professional journal *American Psychologist* in 1966,

"The psychologist owes it to himself to determine not only whether an association exists between two variables—an association which may often be so small as to be trivial—but also to determine the probable magnitude of the association."[72] Dunnette's statement was quoted in one of my methodology textbooks as part of a campaign by the authors to convince researchers to report strength-of-association statistics in order to avoid being "meretricious and deceptive."[73]

Gould implies in his review that Herrnstein and Murray deliberately made a meretricious and deceptive decision to avoid putting the r-squared statistics into the body of their book. Whether or not his charge is true, the fact is that if the two authors had followed good research methodology their book would have been stronger in some ways, but also much shorter. Some of the r-squares in their reported relationships are relatively robust, and deserve to be used as information in debates about public policy. For example, their appendix tells their readers that IQ explains better than a third of the variance in the failure to earn a high school diploma, and more than 31 percent of the variance in white women's probability of going on welfare.[74] But as Gould points out, many of their r-squares are mediocre to anemic. IQ explains less than 2 percent of the variance in the probability of white men being unemployed for at least a month,[75] a bit more than 2 percent of the variance in the probability of being divorced within five years of the first marriage,[76] and about 1.5 percent of the probability of having committed a crime.[77] It explains a tad more than 10 percent of the variance in the probability of living in poverty, a proportion that is substantively significant but not particularly impressive.

Given the weakness of a significant number of their posited relationships between IQ and social outcomes, Herrnstein and Murray's statement that many of the differences in income, unemployment, marriage, illegitimacy, and so forth between African Americans and whites narrow "after controlling for IQ" seems to qualify as political advocacy rather than scientific analysis.[78] This was, of course, exactly Gould's point, and he was right.

Gould wanted his readers to take the evident imperfections in *The Bell Curve* as sufficient reason to reject all of Herrnstein and Murray's arguments. I think this recommended conclusion went too far. There are apparently some genuine relationships between IQ and life chances discussed in the book. By making a single major valid criticism and trying to leverage it into a rejection of the entire analysis, Gould was of course transferring his own political opposition into scientific condemnation. From the perspective of this chapter, the important principle is that, once again, Gould's political purpose was seamlessly connected to his methodological analysis; politics and science were one. Scientists may establish that some groups of humans are superior in important talents, and they may establish that because that superiority is genetic, it implies that inequality is an ineradicable part of social life. But if they do, they will have succeeded in that dubious accomplishment against all of Gould's writings.

The Missing Third Verse

Gould's prominence among the critics of modern intelligence research makes his inattention to the writings of Canadian psychologist J. Phillipe Rushton surprising. Rushton would seem to be the scholar who, above all others, should have inspired Gould to implacable opposition. For, even more than Arthur Jensen, Rushton is the researcher who has lent scientific prestige to theories of mental inequalities that the political left finds anathema. Further, Rushton, unlike Jensen, has explicitly adopted the sociobiological perspective to argue that people of African descent are—there is only one word to describe his position—inferior to both whites and East Asians in intelligence because of differential evolution.

Rushton is not a crank on the margins of scientific discourse. One way of judging the standing of a researcher is to check the number of articles he or she has published in bona fide scholarly journals. By this measure, Rushton is eminent. His articles on race, evolution, and intelligence have appeared in *Personality and Individual Differences*, *Psychological Reports*, *Ethology and Sociobiology*, *Journal of Research in Personality*, *Behavioral and Brain Sciences*, *Intelligence*, *Journal of Personality and Social Psychology*, and others.[79] He has written a book elaborating on his theories published by Transaction, a respectable academic publisher.[80] In 1997 he published a scathing review of the second edition of *The Mismeasure of Man*.[81] Yet Rushton's name does not appear in the indexes to any of Gould's books.

The fact that Rushton seems somehow to have been overlooked by Gould is unfortunate for those who enjoy a good intellectual rumble, for the confrontation of their ideas would have been epic. Rushton's intellectual approach combines the definition of intelligence Gould abhorred with the adaptationist theory of causation that Gould execrated. And the difference in Rushton's use of evidence in the two halves of his theory would have given Gould a good subject for one of his essays.

In regard to the presentation of, and reasoning from, most of his evidence, Rushton is a careful scholar. His book and his articles are meticulous compilations of statistical data showing the differences between East Asians, whites, and "Negroids" on a wide variety of measures. Members of the three groups differ, according to his measures, not only on average IQ but in reaction time to various stimuli, brain size, probability of giving birth to twins, the maturation rate of infants, tendency to develop nearsightedness, skeletal development, age of first pregnancy, aggressiveness, impulsivity, marital stability, law-abidingness, and a variety of other variables.[82] The IQ studies in his publications are subject to all the criticisms published by Gould, Lewontin, and others over the years, and no doubt various scholars have many criticisms of the other correlation studies Rushton has published. My point is not that the conclusions in these studies are correct, but that the research in them is cautiously interpreted, well organized, and based on conscientiously presented, quantified data.

When it comes to the sociobiological theory of evolution underlying the quantitative evidence, however, Rushton abandons scholarly caution. The careful treatment of evidence, characteristic of the other parts of his research, degenerates into lurid speculation. The essence of his theory is that human beings migrating out of Africa during the Pleistocene Age moved into the cold climates of northeast Asia and Europe. Cold climates presented "more challenging environments" than the tropical savannas wherein resided the humans who had stayed home. The migrants were therefore "more stringently selected for intelligence, forward planning, sexual and personal restraint," and a more nurturant parenting strategy than the humans who remained in Africa.[83] These behavior patterns, and accompanying cognitive talents, selected by evolution thousands of years ago, explain the distribution of different talents among groups today.

If Gould and Lewontin had desired a published example of sociobiological research to illustrate their 1979 argument about the flaws in the adaptationist program, they could not have done better than to use this book. Rushton offers no actual evidence to support his theory. In his summary, all of sub-Saharan Africa is hot and all of east Asia and Europe are cold. Survival in cold climates is more cognitively demanding. It mandates more cleverness in gathering food, the manufacture of more sophisticated tools, and the making of fire, none of which are apparently required by living in Africa. Furthermore, the climate of Africa is less predictable than the steppes of Eurasia, which selects for the "r-strategy" of child rearing, in which parents produce as many children as they can while giving less care to each.[84]

Given the scholarly nature of Rushton's analyses of current evidence from survey research, his cavalier treatment of the science of human migration is remarkably disappointing. Human beings did not, as Rushton implies, spurt out of Africa all at once, and all in a northerly direction. Some streams of migrants did indeed flow into Asia, north of the Himalayas, and into China and Siberia, where they certainly encountered a cold climate. But another prominent migration path took the African expatriates and their descendants along the bottom of the Arabian peninsula, around the coastal promontory of India, into the islands of Indonesia and Malaysia, and thence to Australia.[85] Nowhere on this route—even during an ice age—would they have encountered a climate markedly colder than the one they had left behind. By Rushton's logic, therefore, the people who are descended from those who settled along this southern migration route should today exhibit the same low IQs as do Africans. We might expect that he would have tested this hypothesis by analyzing the intelligence test scores of people from that part of the world. But he ignores the obvious implication of his own analysis, and confines his discussions to the scores of respondents of East Asian background.

Moreover, if we put aside the varieties of climate and topography to which migrants would have had to adapt in Asia, and assume, with Rushton, that they would have encountered conditions that were in general colder than humans who remained in Africa, this sort of argument is still breathtakingly short of evidence. He assumes that living in a cold climate would have required more—that is, rewarded more—brainpower than living in an African climate. But it is easy to make the opposite argument.

Africans encounter many more poisonous snakes than do Europeans and East Asians; avoiding or learning to live alongside these would have required a good deal of intelligence. Similarly, Africans are far more likely to encounter dangerous large mammals, not just obvious examples such as lions and buffalo, but less obvious and still deadly animals such as zebras and hippopotami.[86] People living in tropical areas must adapt to the challenge of living with many more diseases than exist in cold climates, including yellow fever, malaria, sleeping sickness, and other scourges. (Sub-Saharan Africans evolved to manage malaria using a genetic strategy other than the growth of intelligence, but that fact does not mean that all diseases would have been resisted in the same way). Tropical residents must adapt to the presence of a large number of very unpleasant insects, including a variety of dangerous species of ants and bees.

Parts of sub-Saharan Africa are prone to drought. Rushton asserts that the appropriate adaptive response to the uncertainty created by this situation would be to adopt a strategy of breeding as fast as possible, in the evolutionary hope that some of the kids would survive (r-strategy). It would be just as plausible, however, to theorize that unpredictable drought cycles should lead to a strategy of spawning fewer children with whom it would be easier to share the sometimes constricted water supply (K-strategy).

I would not insist that any of these rather casual hypotheses have real value in sociobiological explanation. But neither do Rushton's. Without evidence of the environmental challenges actually faced by northeast Asians and sub-Saharan Africans, putative evolutionary theorizing is simply speculation. In this case, the speculation is such as to reinforce some of the most damaging ethnic stereotypes in American history. Such, I believe, would have been Gould's critique of Rushton, if Gould had ever noticed him.

Conclusion: The Science of Inequality

After surveying the development of the science of intelligence testing, and the leftist critiques of it, what can we say about Jefferson's conviction that there is an aristocracy of talent? My view is that few solid conclusions are warranted. There is a mountain of evidence, and much of it tends to lead to the conclusion that there are substantial differences in intelligence between individuals (almost certainly) and groups (perhaps). But a large swath of this evidence is contradictory and ambiguous. It often rests on a set of questionable methodological assumptions. It is sometimes associated with sociobiological theories that are suspiciously lacking evidential support. Whether or not Gould and his allies are actually right in their critique of the genre, they have introduced enough legitimate doubt into the discussion to convince the rest of us to suspend judgment.

Even given this educated hesitation, I would not stop testing individuals. The ability of a *g* score to predict some important behavior with a tolerable degree of accuracy, and better than any other test, is the best-grounded aspect of the IQ research tradition. All other aspects of the concept of IQ, including its dependence on, or independence from, environmental causes, are too equivocal to serve as the basis for confident decision.

What, then, was Gould's contribution to this nondecision? His main accomplishment was to publicize the uncertainty at the heart of intelligence testing. Because his lucid, charming prose style created a lasting interest in his work, this negative publicity is likely to remain a permanent part of the legacy of such testing. Secondarily, and less admirably, he impugned the integrity of those who had conducted, and were conducting, such research. That accomplishment is less important, but it is part of the bargain with Gould, so it is also likely to last.

notes

1. John Locke, *Second Treatise of Government* (Indianapolis: Hackett, 1980), pp. 8, 19, 21, 29–30, 45–46.

2. Thomas Jefferson, letter to John Adams, October 28, 1813, in Adrienne Koch and William Peden, eds., *The Life and Selected Writings of Thomas Jefferson* (New York: Modern Library, 1944), pp. 632–34.

3. Pierre Joseph Proudhon, "What Is Property?" in Albert Fried and Ronald Sanders, *Socialist Thought: A Documentary History* (Garden City, NY: Doubleday, 1964), pp. 201–203.

4. R. C. Lewontin, Steven Rose, and Leon Kamin, *Not in Our Genes: Biology, Ideology, and Human Nature* (New York: Pantheon, 1984), p. 68.

5. J. Phillipe Rushton and Arthur R. Jensen, "Thirty Years of Research on Race and Cognitive Ability," *Psychology, Public Policy, and the Law* 11, no. 2, (2005): 235.

6. Stephen Jay Gould, *The Flamingo's Smile: Reflections in Natural History* (New York: W. W. Norton, 1985), p. 198.

7. Stephen Jay Gould, *Bully for Brontosaurus: Reflections in Natural History* (New York: W. W. Norton, 1991), p. 91.

8. Stephen Jay Gould, *Dinosaur in a Haystack: Reflections in Natural History* (New York: Harmony Books, 1995), p. 246.

9. Robert M. Thorndike with David F. Lohman, *A Century of Ability Testing* (Chicago: Riverdale, 1990), p. 28.

10. Ibid., pp. 64–69, 74–75; Richard Herrnstein, "IQ," *Atlantic Monthly*, September 1971, p. 7.

11. Thorndike and Lohman, *Century of Ability Testing*, p. 72.

12. Robert J. Sternberg and Douglas K. Detterman, *What Is Intelligence? Contemporary Viewpoints on Its Nature and Definition* (Norwood, NJ: Ablex, 1986), p. 3.

13. Mark Snyderman and Stanley Rothman, *The IQ Controversy: The Media and Public Policy* (New Brunswick, NJ: Transaction, 1988), p. 69.

14. Veneta Bastian, Nicholas R. Burns, and Ted Nettelbeck, "Emotional Intelligence Predicts Life Skills, but Not as Well as Personality and Cognitive Abilities," *Personality and Individual Differences* 39, no. 6 (October 2004): 1135–45.

15. Susanne Weis and Heinz-Martin Sub, "Reviving the Search for Social Intelligence—A Multitrait-Multimethod Study of Its Structure and Construct Validity," *Personality and Individual Differences* 42, no. 1 (January 2007): 3–14.

16. Naomi Friedman, Akira Miyake, Robin P. Corley, Susan E. Young, John C. DeFries, and John K. Hewitt, "Not All Executive Functions Are Related to Intelligence," *Psychological Science* 17, no. 2 (February, 2006): 3; Thorndike and Lohman, *Ability Testing*, pp. 111, 113.

17. Snyderman and Rothman, *IQ Controversy*, p. 71.

18. Ibid.

19. Arthur R. Jensen, "Intelligence: 'Definition,' Measurement, and Future Research," in Sternberg and Detterman, *What Is Intelligence?* p. 111; Thorndike and Lohman, *Ability Testing*, pp. 107–108; Herrnstein, "IQ," p. 11.

20. Kaia Laidra, Helle Pullmann, and Juri Allik, "Personality and Intelligence as Predictors of Academic Achievement: A Cross-Sectional Study from Elementary to Secondary School," *Personality and Individual Differences* 42, no. 3 (February 2007): 441–51; Rushton and Jensen, "Thirty Years of Research," pp. 13, 15.

21. Thorndike and Lohman, *Ability Testing*, pp. 120–21.

22. Quoted in Stephen Jay Gould, *The Mismeasure of Man* (New York: W. W. Norton, 1996), p. 21.

23. Rushton and Jensen, "Race and Cognitive Ability," p. 248.

24. John C. Loehlin, "Should We Do Research on Race Differences in Intelligence?" *Intelligence* 16 (1992): 4.

25. Arthur R. Jenson, "How Much Can We Boost IQ and Scholastic Achievement?" *Harvard Educational Review* 33 (1969): 1–123.

26. Philip E. Vernon, *Intelligence: Heredity and Environment* (San Francisco: W. H. Freeman, 1979), p. 198; Snyderman and Rothman, *IQ Controversy*, p. 95; Rushton and Jensen, "Thirty Years of Research," p. 242.

27. Angela L Duckworth and Martin E. P. Seligman, "Self-Discipline Outdoes IQ in Predicting Academic Performance of Adolescents," *Psychological Science* 16, no. 2 (December 2005): 939–44.

28. E. Glenn Schellenberg, "Music Lessons Enhance IQ," *Psychological Science* 15, no. 8 (August 2004): 511–14.

29. William T. Dickens and James R. Flynn, "Black Americans Reduce the Racial IQ Gap: Evidence from Standardization Samples," *Psychological Science* 17, no. 10 (October 2006): 913–20; see also J. Philippe Rushton and Arthur R. Jensen, "Commentary: The Totality of Available Evidence Shows the Race IQ Gap Still Remains," *Psychological Science* 17, no. 10 (October 2006): pp. 921–22; and William T. Dickens and James R. Flynn, "Commentary: Common Ground and Differences," *Psychological Science* 17, no. 10 (October 2006): pp. 923–24.

30. Gould, *Mismeasure*, p. 155.

31. C. Loring Brace and Frank B. Livingstone, "On Creeping Jensenism," in Ashley Montagu, ed., *Race and IQ: Expanded Edition* (New York: Oxford University Press, 1999), pp. 215–16.

32. Thorndike and Lohman, *Ability Testing*, passim; Vernon, *Intelligence*, passim.

33. Lewontin, Rose, and Kamin, *Not in Our Genes*, p. 117.

34. Ibid., p. 243; Elaine Mensh and Harry Mensh, *The IQ Mythology: Class, Race, Gender, and Inequality* (Carbondale: Southern Illinois University Press, 1991), p. 89.

35. Rushton and Jensen, "Race Differences," p. 238.

36. John R. Alford, Carolyn Funk, and John Hibbing, "Are Political Orientations Genetically Transmitted?" *American Political Science Review* 99, no. 2 (May 2005): 153–67.

37. Evan Charney, "Genes and Ideologies," *Perspectives On Politics* 6, no. 2 (June 2008): 303.

38. Ibid, pp. 301–303.

39. Alford, Funk, and Hibbing, "Political Orientations"; Rebecca J. Hannagan and Peter K Hatemi, "The Threat of Genes: A Comment on Evan Charney's 'Genes and Ideologies'," *Perspectives On Politics* 6, no. 2 (June 2008): 329–43.

40. Vernon, *Intelligence*, pp. 202–203.

41. Gould, *Mismeasure*, pp. 21, 28–29.

42. Ibid., p. 29.

43. Ibid., pp. 161, 165, 168, 190, 197–98.

44. Thorndike and Lohman, *Ability Testing*, p. 73.

45. Gould, *Mismeasure*, pp. 245–69, 296–302.

46. Ibid., p. 309.

47. Richard Levins and Richard Lewontin, *The Dialectical Biologist* (Cambridge, MA: Harvard University Press, 1985), p. 159.

48. Thorndike and Lohman, *Century of Ability Testing*, p. 73.

49. John B. Carroll, "What Is Intelligence?" in Sternberg and Detterman, *What*

Is Intelligence? p. 53; Vernon, *Intelligence, Heredity, and Environment*, pp. 202–203; Colin Cooper, *Intelligence and Abilities* (London: Routledge, 1999), p. 129; Rushton and Jensen, "Thirty Years of Research," pp. 235, 238.

50. Mensh and Mensh, *IQ Mythology*, pp. 16–17.

51. Steve Blinkhorn, "What skullduggery?" *Nature* 296 (April 1982): 506.

52. J. Phillipe Rushton, "Race, Intelligence, and the Brain: The Errors and Omissions of the 'Revised' Edition of S. J. Gould's *The Mismeasure of Man* (1996)," *Personality and Individual Differences* 232, no. 1 (January 1997): 169.

53. Cooper, *Intelligence and Abilities*, p. 57.

54. Stephen Jay Gould, "Jensen's Last Stand," in *An Urchin in the Storm: Essays about Books and Ideas* (New York: W. W. Norton, 1987), pp. 124–44.

55. Cooper, *Intelligence and Abilities*, p. 9. Gould, "Jensen's Last Stand."

56. Gould, *Mismeasure*, p. 40.

57. Richard Lewontin, in conversation with the author by phone, October 31, 2006.

58. Ullica Segerstrale, *Defenders of the Truth: The Sociobiology Debate* (Oxford: Oxford University Press, 2000), pp. 2, 41–42, 77, 120.

59. Richard J. Herrnstein and Charles Murray, *The Bell Curve: Intelligence and Class Structure in American Life* (New York: Free Press, 1994), p. 7.

60. Ibid., p. 119.

61. Ibid., pp. 76, 81, 131, 146, 163, 174, 194–95, 246–47.

62. Ibid., pp. 105.

63. Ibid., pp. 279–80.

64. Ibid., p. 340.

65. Ibid., p. 313.

66. Ibid., p. 448.

67. Stephen Jay Gould, "Curveball," in Steven Fraser, ed., *The Bell Curve Wars: Race, Intelligence, and the Future of America* (New York: HarperCollins, 1995), pp. 11–22.

68. Ibid., pp. 11, 20.

69. Ibid., p. 195.

70. Ibid., pp. 164, 249.

71. Ibid., p. 19.

72. Marvin D. Dunnette, "Fads, Fashions, and Folderol in Psychology," *American Psychologist* 21 (April 1966): 343–52; this passage is quoted with approval in the methodology textbook, Julian L. Simon and Paul Burstein, *Basic Research Methods in Social Science*, 3rd. ed. (New York: Random House, 1985), p. 318.

73. Simon and Burstein, *Basic Research Methods*, p. 318.

74. Herrnstein and Murray, *Bell Curve*, pp. 597, 607.

75. Ibid., p. 599.

76. Ibid., p. 602.

77. Ibid., p. 620.

78. Ibid., p. 317.

79. J. Phillippe Rushton, *Race, Evolution, and Behavior* (New Brunswick, NJ: Transaction, 1995); taken from the "References" section, pp. 306–308.

80. Ibid.

81. Rushton, "Race, Intelligence, and the Brain."

82. See the summaries in Rushton and Jensen, "Thirty Years of Research."

83. Rushton, *Race, Evolution, and Behavior*, p. 7.

84. Ibid, pp. 228–29, 231, 241, 255.

85. Steve Olson, *Mapping Human History: Genes, Race, and Our Common Origins* (Boston: Houghton Mifflin, 2002), pp. 123–36.

86. Jared Diamond, *Guns, Germs, and Steel: The Fates of Human Societies* (New York: W. W. Norton, 1999), pp. 171–72.

Chapter 6

science and religion

For most of American history two important sets of interests, capitalists and Christian social conservatives, were on opposite sides of the political fence. Beginning with the inception of the republic in 1787, capitalists, who in American terms were economic conservatives, were based mainly in the North, and used, in succession, the Federalist, Whig, and Republican parties as their vehicle to influence government. Meanwhile, social conservatives, largely based in the South, dominated the Democratic party into the 1960s.[1] Social conservatives opposed many of the intellectual trends of the late nineteenth and early twentieth centuries, including the growing acceptance of the theory of evolution among the educated portion of the population. But the national expression of antievolution sentiment was suppressed by political realities. The Democratic party was effectively in the minority for much of the period 1859 to 1932. When the Democrats achieved majority status during the 1930s, the party was dominated at its national levels by economic and social liberals. The social conservatives, stifled by the liberalism of the national leadership of their own party and unable to imagine a union with the economic conservatives of the party most Southerners considered a traditional enemy, contented themselves with attempting to impose racial segregation and Christian symbolism within the states of the South. Prior to the 1980s, the spasms of antievolution enthusiasm that occasionally provided amusement for sophisticated Americans, such as the Scopes "monkey trial" of 1925, were local in character, chiefly Southern, and not a threat to the institutions of scientific progress in the North and West.

During the 1960s, however, the kaleidoscopic elements of political influences in the United States began to tumble into a new pattern. The first stimulus was probably the several Supreme Court decisions, beginning in 1962, that, in effect, attempted to impose national secularization on local public schools by forbidding both the reading of Bible passages and the institutionalization of public prayers.[2] Religious conservatives, especially in the South, reacted with outrage to these decisions and have continued to flout and evade them. Indeed, although Southern resistance to court decisions forbidding racial segregation in the schools has received much more publicity, Southern noncompliance with the school-secularization decisions has been far more tenacious and successful.[3] Nevertheless, when added to the multiple assaults on traditional authority structures that roiled American society through the decade—particularly the civil rights movement and the counter-culture—the secularizing court decisions had created a siege mentality among social conservatives by the end of the decade.[4]

The second shock to the conservative mindset was the Supreme Court's 1973 *Roe v. Wade* decision declaring that the choice of whether or not to bear a child is a woman's constitutional right, thus invalidating all state antiabortion laws.[5] It seemed to social conservatives that the two pillars of a moral society—religious belief and sexual restraint—had become the target of an evil liberal conspiracy. Through that decade social conservative leaders played on the fury of their constituents, largely but not entirely in the South, to fashion a political movement that rolled all the modernist intellectual trends since Darwin into one great Satanic plot intending to subvert personal and national morality. Although this conspiracy went by several names, depending on the rhetorical tastes of the speaker, it was generally labeled "secular humanism." As one article in the *Christian Harvest Times* summarized the enemy in 1980:

> To understand humanism is to understand women's liberation, the ERA, gay rights, children's rights, abortion, sex education, the "new" morality, evolution, values clarification, situational ethics, the separation of church and state, the loss of patriotism, and many of the other problems that are tearing America apart today.[6]

To those who truly understood the Satanic enemy, then, evolution was not simply a scientific theory that many found disagreeable or mistaken. It was one thread of a vast web of evil that was suffocating national and personal virtue in preparation for the triumph of the Antichrist. When Lipset and Raab investigated American right-wing extremism in the late 1960s, they found that 47 percent of the people in their sample agreed that the country's morals were "bad and getting worse," but that Southerners were greatly overrepresented among those expressing the opinion.[7] The belief in the declining morals of the country was associated with bigoted attitudes toward African Americans, Catholics, and Jews, and, significantly, with intensity of religious commitment.[8] Lipset and Raab did not directly measure attitudes on evolution, being mainly concerned with ethnic intolerance, but their portrait of the right-wing extremists as people who live in a mental world of "generic simplism," "historical moralism," "conspiracy theory," and belief in "the manipulation of the many by the few," strongly suggests that they were measuring the same social impulse that produced the antievolution movement.[9]

The antievolutionists were crackpots, of course. And crackpot theorists are entertaining as long as they are not in power. But with American crackpots running into the millions, and ready to express their angry frustration with the forces of modernity, it was only a matter of time before some entrepreneurial politician saw the chance to unite social with economic conservatives and thus create an unstoppable force at the ballot box.

The requisite political entrepreneur arrived during the Presidential campaign of 1980, in the person of Ronald Reagan. Former free-market, procapitalist conservatives had always felt uneasy associating with the Bible-thumping moralists of the South. The extreme rationalism of the free-market ideology could not seem to coexist with the aggressive irrationality of the social conservatives. But Reagan felt no inhibitions about embracing both forms of conservatism. Already a darling of the economic conservatives, Reagan cemented the alliance with the social conservatives in August when he addressed a conference of fundamentalist preachers sponsored by the Religious Roundtable. Alluding to the constraints of the federal tax code, which forbid tax-exempt organizations such as churches to endorse candidates for public office, Reagan told the assembled multitude, "I know this is non-partisan, so you can't endorse me. But I want you to know that I endorse you."[10] The thunderous approval of the delegates, and of the thousands of Christian conservatives watching their television sets at home, cemented that coalition of the two conservatisms, and began a new era in American politics. Reagan's victory in the balloting in November, his reelection in 1984, and, still more, the capture of both houses of Congress by the Republican coalition in 1994, enabled the social conservatives to participate in power at the national level.

Although Reagan's public statements tended to emphasize his opposition to abortion, he did not neglect to mention the other threads of the humanist tapestry of evil. He was quite explicit in associating himself with the doubters of evolution. At press conferences he urged that the biblical story of creation be taught in the public schools as an alternative to the theory of evolution, which, he said, was increasingly falling out of favor among scientists.[11] But although the president gave symbolic support to a variety of socially conservative causes, in policy terms Reagan's administration put virtually all its energies into enacting measures to please the economic conservatives. Whereas Reagan convinced Congress to cut taxes on the rich, and whereas his administration was noteworthy for its reluctance to regulate the economy, he sponsored no tangible policies to outlaw abortion, restore prayers to public schools, or forbid the teaching of evolution. The actual effect of his eight years in the White House was not so much to give policy help to the antievolution movement as to supply it with symbolic reassurance and national legitimacy.

It was not long, however, before the movement had a victory in the policy arena. In early 1981 the Arkansas state legislature passed Act 590, a law "to require balanced treatment of creation-science and evolution-science in public schools." In May, the American Civil Liberties Union filed suit in federal court, contending that the law violated the First Amendment to the Constitution by creating an establishment of religion.[12] Stephen Jay Gould, among other scientists and philosophers, testified at this trial for the ACLU.

In his history-of-science essays about creationists of the past, Gould had often penned sympathetic portraits of well-meaning intellects who were drawn into scientifically absurd arguments by personal inclinations and social pressures. In exploring the thought of Louis Agassiz, Darwin's contemporary and the last great scientific holdout against the theory of evolution, Gould had been so sympathetic that he turned Agassiz's story into a tragedy. Even though Agassiz had been "without doubt, the greatest and most influential naturalist of nineteenth-century America," he seemed to be unable to absorb the value of Darwin's formulation because "his mind was indentured to the creationist world view and the idealistic philosophy that he had learned from Europe's great scientists." Consequently, over the course of his life following 1859, "all his students and colleagues became evolutionists," and his friends were "saddened by the intellectual hardening of such a great mind."[13] As a result, "we see abundant signs of psychological distress and deep sadness in Agassiz's last defenses of creationism. No one enjoys being an intellectual pariah, especially when cast in the role of superannuated fuddy-duddy."[14] Far from demeaning Agassiz as a roadblock to scientific progress, Gould made us feel the distress of a good mind that time is passing by.

In discussing late twentieth-century creationists, however, Gould was utterly without sympathy. The creationists up through Agassiz were honest intellects wrestling with new evidence and new ideas. Because they were bona fide scientists, they made "truly great and informative errors."[15] But in Gould's view, the creationists of his own day were antiscience zealots seeking to prey on mass ignorance and fear to reverse both the scientific and political progress the human species had made since Darwin. In Gould's writing about this "pseudoscience," therefore, he did not use the language of empathy, but of political polemic.[16] Creationists, he wrote, followed a "graded trajectory from crank to demagogue."[17] Their movement was merely "the latest episode in the long, sad history of American anti-intellectualism."[18] In particular, they employed their "usual mixture of cynicism and ignorance" in "their clearly willful distortions" of his own work.[19]

Gould's testimony in the Arkansas trial was lacking in rhetorical histrionics, but fit into a legal strategy designed to convince federal judge William Overton that Act 590 violated the US Supreme Court's "Lemon test" for evaluating government action with regard to the "no establishment of religion" clause of the First Amendment to the Constitution.[20] In 1971 the Supreme Court had decreed that to avoid violating the establishment clause in this area, government action had to, first, have a legitimate secular purpose, second, not have the primary effect of either advancing or prohibiting religion, and third, not result in "excessive government entanglement" with religion.[21] As a scientific witness, Gould's task was to demonstrate to the judge that Darwinist evolutionary theory was science and that scientific creationism was not, but was religion masquerading as science.

His testimony was characteristically clear, and organized so as to make the legal point with a minimum of irrelevancy. He pointed out several places in which creation science and Darwinian science each permitted inferences of directly clashing, testable predictions. For example, creation science predicts that, because God created all animals at the same time, fossils from all kinds of animals should be found in rock layers of all ages. In contrast, Darwinist theory predicts that different animals will be found in layers of different ages. In this, as in many other examples, Gould recounted, empirical evidence resoundingly confirms the Darwinist and falsifies the creationist expectations.

Gould added that, whenever their expectations seemed to be at variance with the evidence, creationists appealed to supernatural provenance, thus revealing their religious intentions. As an example, he quoted from one of the scientific creationist tomes, *The Genesis Flood*, by John Whitcomb and Henry Morris. On page 76, he told the judge, the authors had stated, "The simple fact of the matter is that one cannot have any kind of a Genesis flood without acknowledging the presence of supernatural events."

It was part of the plaintiff's strategy to establish that the creation scientists were deceptive, not only about the nature of their own enterprise, but about the nature of Darwinist biology. Gould offered several examples of creationists' use of partial and selective quotation to distort or reverse the meaning of biological publications. One of the intentions of the creationists seemed to be to create the false impression that biologists themselves were doubtful that some bit of the geological record supported the Darwinist interpretation. Another was to overlook, or to assert the absence of, supporting evidence for Darwinism, such as transitional fossil forms. And, with particular zest, Gould discussed the creation scientists' mischaracterization of, and misapplication of, the theory of punctuated equilibrium. It was not appropriate, he emphasized, to characterize the theory as "an argument born of despair," as the creationists' would have the world believe, but as an effort to improve Darwinist theory at the level of detail. In no sense did his theory offer support for creationism.

There were other scientific, philosophical, and theological witnesses for the plaintiffs, making basically the same case as Gould from the perspectives of their own expertise. Given the obvious intentions behind the law, they were enough. In early 1982 Judge Overton ruled that Act 590 was the result of a "religious crusade, coupled with a desire to conceal this fact," and therefore unconstitutional.[22] There would be other court cases in the coming decades, but Gould did not participate in them.

All the court decisions went against the creationists. But the crusade, emboldened by Reagan's election, did not cease its attempts to establish the book of Genesis as a scientific authority on a par with *On the Origin of Species*.

Besides the Republican majority, the antievolutionists had another ally on their side after 1980—public opinion. It is always a mistake to invest too much faith in the expressed views of many ordinary citizens. After decades of scholarly research, many experts caution that the distribution of opinion is very sensitive to small differences in the wording of questions. Moreover, a significant proportion of the public cannot be said to have opinions at all, so ignorant and unstable are their views.[23]

With those cautions in mind, however, surveys show that an indeterminately large portion of the American public would prefer that the story of the origin of the universe found in Genesis be the official public creation myth rather than the one that derives from Darwin. In a *USA Today*/Gallup Poll administered to a national sample in June 2007, two-thirds of the public endorsed "the idea that God created human beings pretty much in their present form at one time within the last 10,000 years." When we see that 53 percent in the same poll professed to agree with the statement that "human beings developed over millions of years from less advanced forms of life," it is clear that about a fifth of the public is confused, or deeply stupid, or were drunk at the time they answered the survey. Nevertheless, other survey results suggest that the antievolution majority is not a chimera. A *Newsweek* poll from January 2007 reported that, in regard to the "origin and development of human beings," 48 percent endorsed the idea that God created *Homo sapiens* in their present form, while an additional 30 percent agreed that God guided the process. Moreover, a 2006 Pew Research Center poll reported that 58 percent of the American public "would generally favor" the teaching of creationism alongside the theory of evolution in public schools, while only 35 percent would oppose it.[24]

The attack on the scientific theory of evolution, then, may be led by a small group of intellectual barbarians, but it is nevertheless broadly representative of public opinion. When Gould and other scientists and philosophers excoriate creationism, defend Darwinism, and urge that creationism be kept out of public-school biology classes, they are speaking for an opinion that is, generally speaking, a minority position within American society. That fact, of course, does not make their position incorrect. It does, however, tend to embolden their intellectual adversaries.

Barbarians at the Gates

Not all Christians, nor all Protestant Christians, nor even all creationist Protestant Christians, are of one mind about the theory of evolution that derives from Darwin. In a philosophical sense, it is not even clear that Christians as a group are creationists. Although the creation story in the Old Testament is a rendition about the supposed facts of natural origins, it can be, and often is, interpreted in an allegorical sense. Thus, there are many examples of Christian intellectuals who accept the scientific theory of evolution while continuing to draw moral inspiration from the Bible.[25] Moreover, not all scientists are hostile to religion in general or to Christianity in particular. Whereas it is generally true that natural scientists are a secular lot, and that the more eminent the scientist, the more likely he or she is to profess atheism, the tendency does not express a unanimity.[26] Therefore, there is no struggle between science and religion taking place in modern society. There is a culture war, but its composition cannot be accurately described as consisting of religious antievolutionists on one side and atheist scientists on the other.

Moreover, even among the large group of Christian conservatives who disbelieve the theory of evolution, there are differences. Some—"young earth" creationists—are Biblical fundamentalists and therefore reject most of modern science.[27] Others accept most of science, and even parts of the theory of evolution. Among this latter group, the most interesting and most vocal in recent decades have been those who, in general, do not repudiate the idea that Darwinian processes fashion microevolutionary variation in species, but maintain that only "intelligent design" can explain macroevolutionary patterns of life's development, especially speciation. Like the "scientific creationists" of a generation earlier, intelligent-design proponents insist that they are scientists, not theologians, and although the claim is untrue they have managed to make some headway among the public over the last two decades. Nevertheless, the argument about evolution is one in which the overwhelming majority of scientists and scholars in general, and a large proportion of religious thinkers, face a group of very conservative and generally fundamentalist Protestants allied with a tiny minority of scientists and other scholars, and further allied with a large but indistinct group of the ordinary, poorly informed citizenry.

Modern creationists as a group understand that they cannot simply thump Bibles and call the faithful to worship; they must meet scientific arguments with appeals that at least sound scientific to the uninitiated. Since about 1970 they have developed a set of themes upon which they elaborate with ingenuity, to the effect that Darwinian theory and supporting evidence are so unsatisfactory that the theory must be disbelieved as a whole. Implicitly—or, in a surprisingly high percentage of cases given their scientific pretensions, explicitly—they offer the creation myth in the book of Genesis as the plausible alternative to a discredited Darwinism. By now, the shelves of Christian bookstores are laden with this sort of antiapologetics for evolutionary theory.

Periodically, scientists or philosophers write books or articles exposing the sophistry, mischaracterization, and abuse of evidence offered in these books.[28] Scholars point out that over the last forty years or so a huge mass of biochemical evidence has accumulated, virtually all of it consistent with the interpretation that all living things are related, and that all are descended from a common ancestor. They point out that the creationists tend not to offer evidence for their own ideas about how the earth and its life came to be, but to limit themselves to criticizing evolutionary theory, falsely assuming that the story in the book of Genesis must become the default interpretation if Darwinism is discredited. Scholars point out, however, that there is an almost infinite number of creation stories, all of which are supported by as much evidence—none—as the Christian fundamentalist story. They further remind us that the fundamentalists' implied causal argument to replace Darwinism—God did it—is not an explanation but an invitation to a worship service. They inform us that science, including evolutionary science, corrects itself, offering many examples of faulty biological ideas that are no longer believed because the evidence proved them wrong, and ask us to recall that creationists brag about the fact that their understanding of life has not changed in two thousand years.

These refutations do not cause the creationist writers to change their opinions, of course. Many of the creationists learn from the criticisms, however, and thereby fashion even more inventive casuistries in their next book. In fact, the label *intelligent design* was adopted by a new group of religious advocates after the defeat of the creation science movement in the courtrooms of the 1980s.

I do not intend to rework the ground by supplying yet another systematic refutation of the arguments of creationists. My purpose in this section is twofold: first, to examine and evaluate a few of the most interesting arguments that creationists make against evolutionary theory, and second, to analyze the way creationists make use of Gould's own writings to come to conclusions that Gould abhorred.

The first interesting creationist claim is that modern evolutionary biologists, for all their input of effort and resources, have not been able to produce a single new species in the laboratory, or observe a single new species being born in nature.[29] As James Barham summarized the creationist position in 2004, "Since we cannot produce new species at will, we are in no position to state categorically that natural selection is the driving force behind evolution."[30] Or as Ann Coulter slightly restated the point in 2007, "Human breeders have not been able to produce one biologically novel structure in the laboratory."[31]

On the face of it, this is a rather unfair criticism to make of evolutionary biologists. Since Darwin, most theorists have emphasized the extremely slow pace of evolution, with species transforming into other species over tens of thousands or millions of years. Even Eldredge and Gould maintained that evolutionary punctuations that produced new species in *geologically* rapid time scales lasted thousands of years, far longer than a human lifetime. When Richard Goldschmidt theorized during the 1940s that speciation might occur in a single generation, his thinking was firmly rejected by other biologists.[32]

Most creationists, however, would probably prefer that Goldschmidt had been correct. They would want species to have come into existence in an instant. The notion of slow transformations is anathema to the worldview that expects "kinds" to have been created by God through magical and presumably very rapid means. If anyone could come up with a species that arrived on the scene in a Goldschmidtian manner, it would be interpreted by creationists as a refutation of Darwin and a victory for their side. Indeed, some creationists, misunderstanding the idea of punctuated equilibrium to be a theory of *saltationism*, or instantaneous appearance, hail it as "the one evolutionary theory that tries to remain true to the record."[33] They demand that biologists come up with the very sort of evidence that would support the creationist argument.

Such evidence will not be forthcoming. As a matter of fact, however, there are both laboratory experiments and field observations that provide about as much information as could be expected in the length of a few human lifetimes concerning the process of speciation. In the laboratory, for example, there have been a variety of cross-breeding experiments in which biologists take a sample of rapidly reproducing creatures (generally, fruit flies), isolate them into two groups that differ in some morphological or behavioral trait, allow the separated groups to breed in isolation for up to one hundred generations, then remix all the individuals and see if sexual isolation has been established. Since breeding isolation is the first sign that two groups are separating into different species, the biologists assume, not unreasonably, that if they create it they have begun the process of fashioning two species out of one. Although these studies have produced different strengths of sexual isolation, and although some of them are subject to various kinds of methodological criticisms, the basic conclusion to be learned from them is that it is possible to begin the process of speciation in the laboratory.[34] In one experiment in 1966, appreciable sexual isolation emerged after only five generations of "divergent selection."[35]

There is also information from field studies. It is possible, by allowing some reasonable inferences from observable processes, to confirm the notion that species are transforming into other species in their natural state, before our eyes. The best example is probably the apple maggot fly, *Rhagoletis pomonella*.[36] In the Hudson River valley, this species used to live and breed exclusively on the hawthorn, a shrub of the rose family. In the mid-nineteenth century, apple trees were introduced into the valley, and farmers soon discovered that some of the flies had laid their eggs on apple trees. Subsequently the maggots became a serious pest in East Coast orchards. Because hawthorns and apple trees flower at different times of the year, the two races of the one species now not only eat and mate on different plants, but during different seasons. Lab experiments have shown that individuals of the two races can mate and produce fertile offspring. Nevertheless, because of their different life experiences the hawthorn-infesting flies now mate exclusively with other hawthorn-infesting flies, and likewise for the flies infesting apple trees. Genetic analysis has established that the two races differ in the frequency of certain proteins coded by their respective DNA. In sum, while the two races once consisted of a single interbreeding gene pool, now they consist of two slightly different gene pools whose individuals are no longer interbreeding. It would be a plausible expectation, therefore, that the two gene pools will continue to diverge as the two races continue to live separate lives on different plants. The two races of this fly appear to be in the process of becoming distinct species.

When creationists ask to be shown the actual process of speciation in action, therefore, it is an unreasonable request. Yet only a slightly different demand can be satisfied by drawing a few plausible inferences from evidence available to everyone. While creationists cannot be expected to accept the apple maggot fly as an instance of visible speciation, or the results of divergent selection experiments by biologists, the rest of us might consider whether their accusation that no examples of ongoing speciation exist is an accurate characterization of nature.

The second interesting criticism made by creationists, and the one that occupies the largest part of almost every book and article penned by people attempting to discredit Darwinism, is the alleged lack of transitional forms between species in the fossil record. Some creationists demand "hundreds or even thousands of different species, all unambiguously intermediate in terms of their overall biology and in the physiology and anatomy of all their organ systems."[37] But, given the unsatisfactory nature of the fossil record, this is plainly an unrealistic standard. More reasonable creationists understand, and expect their readers to understand, that many transitional types of organisms will be missing from the record. They therefore attempt to convince their readers that *no* acceptable transitional forms have ever been found, or ever will be found, because they do not exist.

When creationists make this type of argument, they come perilously close to offering a testable hypothesis. If the prediction from creationist thought is that intermediate forms never will be unearthed because they have never existed, creationists are somewhat inadvertently sketching out a critical test between Darwin's expectations and theirs. In reality, anyone who has ever read a smidgen of creationist literature knows that no amount of falsified predictions would convince creationists that their position was mistaken. Creationists do not dwell in a psychic world where opinions must be molded to evidence. Nevertheless, for the rest of us it might be a helpful exercise to put the creationist hypothesis to the test.

One of the creationists' favorite types of animal for illustrating the lack of transitions is the whale (of which, of course, there are several living species). As Michael Denton wrote in 1986, there were are no examples of fossils to illustrate the transition between land and aquatic mammals:

> Taking into account all the modifications necessary to convert a land mammal into a whale—forelimb modifications, the evolution of tail flukes,

> the streamlining, reduction of hindlimbs, modification of skull to bring nostrils to the top of head, modification of trachea, modification of behavior pattern . . . one is inclined to think in terms of possibly hundreds, even thousands, of transitional species on the most direct path between a hypothetical land ancestor and the common ancestor of modern whales.[38]

At one level, this is clearly an unreasonable request. We could never, for instance, find examples of fossilized modifications of "behavior patterns." But on another level, it is close enough to a testable proposition to inspire a search for evidence. While the expectation of hundreds of transitional species is, given the problems of fossils, not likely, a fair-minded person might recast Denton's rhetoric as the prediction that no intermediate species, or protowhales, will ever be found. Therefore, if even a fairly small number of transitional forms—say, five—were discovered after Denton's book appeared in 1986, the theory of evolution would be rather strongly confirmed and the theory of supernatural creation rather strongly disconfirmed.

As it happened, about the time Denton was writing, paleontologists were finding a partial fossil of the oldest whale in river sediments bordering an ancient lake in Pakistan. Then, in 1990, 1993, and 1994, four more partial skeletons were unearthed, each providing a further example of a creature more advanced on the road to whaledom. Of one of these newly discovered species, *Ambulocetus natans*, Gould wrote that it was "so close to our expectations for a transitional form that its discoverers could not provide a professional paleontologist with the greatest of all pleasures in science—surprise."[39] Like modern whales, this creature swam by undulating its flexible spine up and down. Like seals and unlike whales, however, its main propulsive force was provided by its large hind feet. Like both types of animals, its forelimbs were so small that they would not have been useful in propulsion through the water; they were probably used for steering and balance. Gould quotes the authors of the scientific report in stating that *Ambulocetus* "represents a critical intermediate between land mammals and marine cetaceans" (whales).[40] From the available evidence, this statement would appear accurate. The creationists' hypothesis, if it can be called that, is thus disconfirmed.

The third interesting creationist criticism is that Darwinism lacks a "theory of the generative."[41] Darwinians posit that mutations occur in the genome during reproduction, thus leading to various kinds of changes in the embryo, depending on where they are located. It is emphatically a major part of the theory that such mutations are random, that is, that they do not occur in response to the organism's need for them.[42] They happen because of some sort of mistake in the DNA copying mechanism, caused by heat, injury, cosmic rays, or various other perturbations in the environment. Whether a given mutation creates a beneficial change in the organism is entirely fortuitous, and as a general rule the vast majority of mutations will be deleterious and therefore lethal. Evolutionary theorists insist, however, that such mutations are so numerous that, given the fact that even an infinitesimally small percentage of them are helpful rather than fatal to the offspring, they can account for all evolutionary change. As Philip Kitcher summarized the theory:

> It is clear that the appropriate mutation rates to be considered are *population* rates of mutation per generation. . . . Assuming a population of 1 million organisms, and standard mutation rates per zygote [a fertilized cell], a conservative estimate of the number of mutations per generation would be on the order of 100,000 mutations per generation. . . . Let us suppose that 1 in 1000 of these are advantageous in the population's environment. That allows for 100 advantageous mutations per generation. Given 25,000,000 generations, it is hardly implausible that 1,000,000 mutations would become fixed in the final population.[43]

The way Kitcher puts the case, it is not implausible. But as creationists argue in virtually every one of their publications, animals do not actually evolve one single step at a time. They must evolve many coordinated steps at a time, or the resulting creature would be an unviable monster. As Denton emphasized, to go from a hippopotamuslike land grazer to a modern whale involves multiple steps on many different bodily systems. The skeleton must be radically modified. The nostrils must move from the snout to the top of the head. The teeth, appropriate for cutting and grinding grass, must change into baleen, appropriate for seining krill. The usual mammalian layer of fat under the skin must become very much thicker—that is, change into blubber. The respiratory system must change to one that can function under the extreme pressures of a dive; likewise the circulatory system. The coordination of tongue, mouth, and lungs that produces moos and grunts through the air must change to one that produces very different sounds to travel through the water; similarly the ears and the sections of the brain that process sound must be modified so as to be sensitive to a kind of sonar.

Most importantly, all these changes must be coordinated. A protowhale must evolve a cooperating system of baleen, blubber, and sonar, not one of baleen, feathers, and yodeling. The method of calculating overall probabilities, therefore, cannot be one like Kitcher's, in which individual probabilities are estimated sequentially. The tiny probability of producing a useful mutation in the dentition must be multiplied by the tiny probability of producing one simultaneously in the circulatory system, the tiny probability of producing one simultaneously in the skeleton, and so on. Furthermore, if such an astronomically improbable concatenation of mutations were ever to come about, the result would only be one miniscule step on the road to a whale. The next step would be equally improbable. The actual probabilities, and the total number of steps that would have to come about, are all unknown. But on a scale of general plausibility they begin to look very close to flatly impossible. The argument made by virtually all creationists is that once we take the necessity of coordinated random mutations into account, the Darwinist scenario becomes wholly unconvincing. The only way billions of such transformations could have come about is if there is some sort of guiding force within, or outside of, the genome to make sure that the right mutations come along in the right embryological systems at the right time.[44]

This alone of the creationists' attacks on Darwinist theory has some plausibility. Evolutionary biologists have not given enough attention to the origin of mutations, and so their account of the mainsprings of life has a genuine vulnerability. Creationists exploit the weakness for their own antiscientific purposes, but they did not create it.

Moreover, creationists are not the only ones who have come to the conclusion that the probabilities necessary to make a theory based on undirected but coordinated random mutations work are not possible. Arthur Koestler, who might be termed one of the major freelance intellectuals of the twentieth century, wrote a series of books criticizing the basic ideas of Darwinist theory.[45] An atheist, Koestler had no truck with creationism. He believed that the key to evolution would have to be found either in *Lamarckism* (the doctrine that evolution proceeds through "directed variability," an "intrinsic bias in adaptive directions"[46]) or *vitalism* (the doctrine that there is in life itself a knowledge of its own needs, and thus that organisms know at an unconscious level in what direction they need to mutate). Darwinists repudiate both doctrines, and there is very little evidence that either theory might account for the direction of evolution. Koestler's main energies were devoted to criticizing Darwinism, and when he tried to develop evidence for Lamarckism, as in *The Case of the Midwife Toad*, he fumbled the job. But he was a perceptive critic, not a biologist. As with the creationists, his writings do not explain how life develops but they do provide reasons for doubting that the Darwinist account is complete.

When faced with this sort of probabilistic skepticism, Darwinists typically refer to decades of research and theory on *pleiotropy* (the occurrence of genes that affect two or more aspects of the organism that are apparently unrelated), *epistasis* (an interaction in which one gene interferes with the expression of another gene), and *homeotic genes* (regulatory genes that control the development and expression of other genes that actually create the structures and behaviors of the organism).[47] Without a doubt, these developments together create a portrait of evolutionary change that is a good deal more flexible and prolific than one based on the idea that genes change one at a time. It is clear that a mutation anywhere on the genome can have multiple and far-reaching effects on the embryo, and thence on the organism. Nevertheless, no matter how easily natural selection can convert a mutation into several different types of phyletic change, there still remains the problem of the origin of the change itself. The expectation that mutations must have come along that permitted the environment to select a series of ever-more-whalelike, or ever-more-froglike, or ever-more-ostrichlike combinations of bodies and behaviors is a conviction that seems, to this secular nonbiologist, to need more supporting evidence and theoretical elaboration.

The skepticism of nonbiologists such as the creationists and Koestler would be of interest only for its literary value if they were not joined in their doubts by some biologists. But among the students of life there has been, ever since Darwin, a minority undercurrent of dissatisfaction with an explanatory framework that relies on coordinated chance mutations. Goldschmidt and Gould were part of this undercurrent. As Gould wrote in his final book, "The central question of evolutionary theory remains: what creates the fit?"[48] His own answer was that "small changes early in embryology accumulate through growth to yield profound differences among adults."[49] For all his eccentricity, therefore, he remained, at his core, an orthodox Darwinist. But his willingness to question the assumptions of the Darwinist mainstream around the margins of the theory provided non-Darwinists with the ammunition they desired to take potshots at its heart.

Gould was twice a blessing for creationists. First, he expressed trenchant criticisms of many aspects of the Darwinian edifice. Second, his lucid prose style permitted nonspecialists to at least partially understand the issues that were being debated within biology, and therefore assimilate his criticisms of the orthodoxy to their own uses.

Virtually all creationists used (and still use) the theory of punctuated equilibrium, especially Gould's statement in one of his popular essays that the lack of transitions in the fossil record was the "trade secret" of paleontology. They like to portray the theory as a last-ditch attempt to paper over the lack of evidence for evolution. "Punctuated equilibrium," crows Philip Johnson, attempts to salvage Darwinism "by making the process of change inherently invisible. You can imagine those peripheral isolates changing as much and as fast as you like, because no one will ever see them."[50] As Ann Coulter puts it, Eldredge and Gould's theory requires "a more sophisticated supernatural occurrence," than does a theory of intelligent design, and therefore is embraced only by those who are disposed to believe in a "nontheological miracle."[51]

Creationists also like to reinterpret Gould's argument in *Wonderful Life* that the evolution of humans was radically improbable, and that if the tape of life were to be rewound and allowed to unspool again, it is unlikely that something resembling *Homo sapiens* would be the end result. "Gould's argument is powerful," agrees Kenneth Poppe. "Without present goals, 'recapitulation' of ultra-complex structures, such as eyes, would be ultra-fantastic occurences. . . . Gould was absolutely right. Evolution needs a more credible explanation." And Poppe, of course, is happy to tell us what that explanation is: "God fashioned all these things from an incomprehensible source completely beyond the reach of science."[52]

Creationists also enjoy trying to refute the thesis of one of Gould's better-known popular essays. Traditionally, Gould told us, arguments in favor of evolution emphasized the marvelous perfection with which evolution could adapt the structures of any organism to that organism's needs. But this reliance on "optimal design" for proofs of evolution, he argued, was a mistake.[53] Perfect design is at least as good an argument for supernatural creationism as for evolution through natural means. After all, an omniscient and omnipotent creator would be likely to fashion everything perfectly. Evolution, on the other hand, because it must work with whatever materials are available, is likely to cobble together organisms in a way that is not perfectly efficient, but only the best option under the given circumstances. "Odd arrangements and funny solutions are the proof of evolution—paths that a sensible God would never tread but that a natural process, constrained by history, follows perforce."[54]

Consider, he offered, the "thumb" of the giant panda. The panda has a paw with five digits, like its relatives, true bears. But bears also have an enlarged radial sesamoid bone—the slight bumpy outcrop just below the thumb on the wrist of a human hand. In the panda, this bone has become even larger and has been supplied with a muscle of its own to make it into a sixth digit that functions somewhat like a real thumb for grasping. This unique sixth-digit thumb is not very efficient, but it permits the panda to hold the shoots of its only food, bamboo. "An engineer's best solution is debarred by history," Gould informs us. "So the panda must use parts on hand and settle for an enlarged wrist bone and a somewhat clumsy, but quite workable, solution."[55] This jerry-built mechanism would make no sense as the outcome of God's perfect plan, but it is completely understandable as the consequence of evolution building by small steps with the materials available. Thus, "Nature is . . . an excellent tinkerer, not a divine artificer."[56] The imperfection of nature's creatures proves that they are not the result of a supernatural designer.

Wrong, wrong, say the creationists. Gould must give us a model of perfect design and show how far the panda's thumb diverges from it. Otherwise, his depiction of this particular example of God's handiwork is arbitrary and therefore irrelevant. "Not knowing the objectives of the designer," asserts Michael Dembski, "Gould is in no position to say whether the designer has come up with a faulty compromise among those objectives."[57] Paul Nelson concurs: "The panda's thumb is a sign of history—i.e. of descent—only when one is certain that a 'sensible God' . . . would not stoop directly to contrive such oddities."[58]

This argument is so audacious that it approaches chutzpah. The goal of the advocates of intelligent design, after all, is to establish that the arrangement of the parts of creatures is, well, intelligent. Gould's assertion that the contrivance of the panda's front paw was not particularly intelligent, in the sense of being useless for most tasks but just barely good enough for one, would seem to be directly germane to their purposes. For intelligent design advocates to recoil from the analysis and claim that we are not allowed to discuss the intelligence of given designs because we cannot know the objectives of the designer would seem to be a repudiation of their own proclaimed purpose. Either they are arguing that organisms show evidence of having been fashioned by a purposeful intellect, in which case we must be allowed to judge that evidence, or they are arguing that we are not allowed to investigate the purposes of the creator, in which case their claim that they are inquiring into intelligent design is a ruse.

Gould had no doubts about which description applied to the modern creationists, and no hesitation about expressing his opinion of their expropriation of his writings. Although he could be rough and even rude when engaged in polemics with other biologists, he rarely stooped to name calling, and when he did, the labels he placed on ideological adversaries were refined, like "ultraorthodox." But he was completely unrestrained when characterizing creationists. They were "modern Yahoos,"[59] a "philistine scourge,"[60] who engaged in "pseudoscientific practices as illustrated by their clearly willful distortions" of his own ideas.[61] The reason for their deceptions was clear to Gould. Creationists are essentially "right-wing evangelicals who advance the literalism of Genesis as just one item in a comprehensive political program that would also ban abortion and return old-fashioned patriarchy under the guise of saving American families. Political programs demand political responses."[62]

Unlike many biologists, Gould understood that the creationist challenge could not be met only at the level of biological discussion. If political programs mandated political responses, then discussion that was explicitly political had to be part of biology's response. But what, exactly, could a biologist with rhetorical skills do to make a political difference in the culture wars?

Salvation

Because of the great success of his popular essays on evolution, Gould had a large audience for any idea he wanted to float. Toward the end of the 1990s he apparently became concerned with the growing intensity of the culture war in the United States, and decided to try his hand as a mediator on a grand scale. In 1999 appeared *Rocks of Ages*, his proposal for calming the conflict between science and religion.

Anyone who, remembering Gould's frequent insults of creationists over the years and the absolutely secular tone of all his essays, expected him to make an attack on religion, was surprised. "I speak of the supposed conflict between science and religion," he wrote in the first paragraph, "a debate that exists only in people's minds and social practices."[63]

> Science tries to document the factual nature of the natural world, and to develop theories that coordinate and explain those facts. Religion, on the other hand, operates in the equally important, but utterly different, realm of human purposes, meanings, and values—subjects that the factual domain of science might illuminate, but can never solve. . . . I propose that we encapsulate this central principle of respectful noninterference . . . by enunciating the principle of NOMA, or Non-Overlapping Magisteria."[64]

The *magisterium*, or intellectual domain, of science, would be recognized as supreme in matters of fact. The magisterium of religion would be recognized as supreme in questions of "ultimate meaning and moral value. These two magisteria do not overlap."[65] Specialists in the two fields should not stop talking with one another, but each should recognize the superiority of the other in the others' magisterium. Further, each would recognize that there is no conflict between their two fields. Both scientists and the great majority of religious leaders would unite against creationists, who represent an "intellectually marginal and demographically minority view of religion that they long to impose upon the entire world."[66] Therefore, scientists should realize that "the enemy is not religion but dogmatism and intolerance," and should unite with responsible religious leaders to repudiate the extremists.[67] Thus mutually respectful and supportive, science and responsible religion can each proceed to enlighten the world in their own distinctive ways.

At the intellectual level, there must be two responses to this suggestion. The first is to remind readers that much of this book has been an exploration of the way that Gould spent a large part of his career drawing political implications from his scientific theories, and scientific implications from his political values. If there was one American scientist in the twentieth century who mixed the magisteria of fact, morality, and ultimate meaning in his own work, it was Gould. For him to turn around and recommend the separation of the two spheres begs for some sort of explanation.

The second response is to point out that all religious doctrines must of necessity make claims about factual reality. Religions can and often do posit that there is a spiritual reality on a different plane from factual reality, but religions do not, as a rule, teach that factual reality does not exist and can therefore be ignored. Buddhism teaches that the mundane plane of physical life is an illusion to be overcome, but it admits that the illusion is powerful and that the overcoming therefore requires strenuous effort. The only way that Judaism, Christianity, and Islam could follow the prescription of staying in their own magisteria would be to proclaim that every one of their beliefs—including the belief in God—was simply allegorical, thus relegating their central tenets to the realm of inspirational fiction.

By recommending this mutually respectful withdrawal into separate spheres, Gould did something that would, before he wrote, have been judged impossible—he united scientific atheists and fundamentalist Christians in vociferous agreement with each other. The members of each group saw without much trouble that both religion and science make incompatible claims about the real world. It might be possible for scientists to abjure all moral claims—in fact, many scientists do so. But it would never be possible for religion to renounce its claim to speak about the real world and remain anything but a shriveled and vestigial institution. As Richard Dawkins, the most outspoken atheist in biology, pointed out, Gould's suggestion "founders on the undeniable fact that religions still make claims about the world which, on analysis, turn out to be scientific claims."[68] Any assertion that there is such a thing as heaven or hell; or a gargantuan invisible magician who created the world and listens to our prayers; or that there once was a man who was born of a virgin, who was at the age of thirty-three executed, but resurrected three days later; or any one of many other such claims, is, if not wholly symbolic, then a scientific hypothesis. Furthermore, Dawkins wondered, "Does Gould really want to cede to *religion* the right to tell us what is good and what is bad? The fact that it has nothing *else* to contribute to human wisdom is no reason to hand religion a free license to tell us what to do."[69] Meanwhile, for perhaps the first and only time, creationist Stephen Meyer found himself agreeing with Dawkins:

> To make NOMA work, its advocates have to water down science or faith, or both. Certainly Gould did—he said religion is just a matter of ethical teaching, comfort, or metaphysical beliefs about meaning. But Christianity certainly claims to be more than that. . . . There might be some religions that can fit comfortably with NOMA. But Biblical Christianity—because it's built not just on faith, but on facts—simply cannot.[70]

At the level of intellectual consistency, Dawkins, Meyer, and the many on both sides who have expressed the same opinion, are quite right. The two different explanations of how life came to be the way it is on earth, the traditional Christian and the Darwinian, cannot both be right. Observed purely from the perspective of reason, it is impossible not to agree with Dawkins when he remarks, "I simply do not believe that Gould could possibly have meant much of what he wrote in *Rocks of Ages*."[71] Because I agree with Dawkins on this point, I conclude that Gould, who had proven many times that he was willing to challenge received opinion and stand up to majorities, must have had other purposes than intellectual consistency when he proposed NOMA.

In science, of course, intellectual inconsistency is a forgivable imperfection, but nevertheless a mistake that all intellectuals are bound to try to correct. To be deliberately inconsistent would go against the temperament and training of any scientist. In democratic politics, however, intellectual inconsistency is a major, if infrequently praised, virtue. As recent world events have often reminded us, social peace is not to be taken for granted by anyone thinking about how to govern a country. Some scientists argue that violence is an evolved characteristic of human beings.[72] Others, including Gould, argue that both violence and harmony are functions of the historical situation.[73] Whichever causal theory is correct, internecine murder is common enough to concern anyone who thinks seriously about structuring a successful society. And, as recent history has emphasized, religious disagreement is a very common cause of social upheaval. Any strategy to blur the ideological differences between religious sects, or between religion and secularism, is probably a contribution to social peace.

When politicians in democratic societies go about their business of aggregating support, they must put together coalitions of groups whose members, if they were to have a serious conversation with members of other groups, would probably discover that they have alarming disagreements about values and principles. A national community, when looked at from the point of view of the beliefs and values of its citizens, is a collection of cliques at odds with one another. In order to attract a majority coalition, politicians do not attempt to clarify the differences between groups; quite the contrary. One of the democratic arts is to obfuscate conflicting ideas and smooth over lifestyle differences so that people can live together with at least minimum levels of harmony. A democratic strategy, then, is to deliberately create public ambiguity. As political scientist Benjamin Page has reported, in American politics "the most striking feature of candidates' rhetoric about policy is its extreme vagueness. . . . Even incumbent presidents . . . are surrounded by a cloud of vagueness"; politicians "make an art of ambiguity."[74] Page himself does not think that such ambiguity is a good thing, but I find it more defensible. In the seventeenth century, in the midst of the English civil wars, Thomas Hobbes reminded all subsequent political thinkers that the first purpose of political philosophy must be the establishment of social peace.[75] If that peace comes at the price of some ambiguity in public discussion, it is a good bargain.

Early in his career, during the 1970s, Gould was quite willing to assault traditional bastions of scientific and political order. By the 1990s, however, he had evidently learned to see the worth of cultivating social peace. In *Rocks of Ages* he discussed the meaning of a word he had come to value: "irenic." Originally "defined in opposition to polemics as a branch of Christian theology that 'presents points of agreement among Christians with a view to the ultimate unity of Christianity' (*Oxford English Dictionary*)," in modern parlance it had acquired a wider meaning. Irenic people and proposals "tend to promote peace, especially in relation to theological and ecclesiastical differences."[76] "Now, I'm an irenic fellow at heart," he assured us, no doubt surprising his long-term readers. "Irenics sure beats the polemics of ill-conceived battle between science and religion."[77]

There is evidence to suggest that this was a good strategy. For one thing, a fairly large number of scientists and religious thinkers have either endorsed NOMA directly or written their own recommendations that closely parallel Gould's ideas.[78] They have been happy to join the coalition of the muddled middle. For another thing, recent events make clear that there are, in the United States at least, a large number of religious citizens who do not want to be pushed into a conflict with science.

A case in point is the citizenry of Dover, Pennsylvania.[79] In 2004, under the influence of some of the advocates of intelligent design, the school board of this small town adopted a policy for its ninth-grade biology classes mandating that before they were exposed to Darwin their teachers should read a statement warning them that the theory of evolution was controversial among scientists, and that there were available alternatives. Eleven parents promptly filed suit in federal court, alleging, like the plaintiffs in the Arkansas suit of twenty-four years earlier, that the policy violated the First

Amendment's guarantee of a separation of church and state. In December 2005, federal district judge John Jones (a Republican appointee), agreed with them and voided the policy.

Of more immediate relevance, however, in the previous month's school-board election all eight members who had instituted the policy were defeated for reelection. Postelection interviews with the victorious insurgent candidates made clear that they were definitely not running to enshrine atheism as school policy. They emphatically declared that they were believing Christians, but that they did not like to see religious dogma replace scientific education in their public schools. These, then were the democratic majority. They were not scientists and they could probably not have analyzed the subtle points at issue between creationists and evolutionary biologists. They were comfortable living with, and perhaps not noticing, the intellectual incoherence of their public stance. They asked for the comfort of religion combined with the products of science, not the rigor of intellectual consistency. These are the people whom both Dawkins and the partisans of intelligent design would like to force to make a choice between the two clarities of atheism and creationism. They are the people whom Gould would have been happy to include in a coalition of the ambiguous. I'm with Gould on this one.

Conclusion

The political/biological conflicts discussed in the earlier chapters of this book were "inside-out" controversies, so to speak. They arose within evolutionary biology, and, vibrating in harmony with various social concerns in the larger world, found interests and individuals within the context of a mass audience willing, even eager, to listen to the points and counter-points of the disputes. Gould was very good at this sort of expansion and transmutation of biological issues. He sometimes played the part of agitator, bringing the quarrels of the small, precise forum of evolutionary biology into the larger and cruder forum of public disputation.

The creationist controversy, however, is an "outside-in" struggle. It is imposed on evolutionary biology by interests and individuals who exist external to the realm of science, and who share neither values nor points of view with scientists. Some scientists seem to enjoy intrascientific dispute, and even those who do not grudgingly accept it as inevitable, recognizing that tolerance of quarreling is part of the price of admission to the guild. But scientists do not enjoy conflict with creationists on any level, for they recognize that the religious zealots are a fundamentally different sort of character from scientists. Science enshrines rationality as an ideal; creationism enshrines irrationality. Thus, scientists quite properly regard creationists as dangerous aliens who threaten the survival of the scientific enterprise itself. Inside-out conflict is part of the game; outside-in conflict is self-preservation, and no fun.

The difference is evident in the personal relations of scientists. Richard Dawkins spent almost thirty years publicly arguing with Gould on practically every point about which it was possible to disagree in evolutionary biology. The two not only exchanged eloquently worded refutations and counter-refutations on a myriad of topics, but always included personal barbs that, although they never quite passed into the realm of vulgar insult, were clearly disdainful putdowns. Yet shortly after Gould died, Dawkins wrote of their relationship, "Although we disagreed about much, we shared much too, including a spellbound delight in the wonders of the natural world, and a passionate conviction that such wonders deserve nothing less than a purely naturalistic explanation."[80]

If Dawkins had died first, it is easily possible to imagine Gould composing a similarly affectionate memorial to him. But it is not possible to imagine either of them writing such an appreciation of a creationist, or vice-versa. In the context of civilized discourse, scientists and creationists are, quite simply, different species.

notes

1. David F. Prindle, *The Paradox of Democratic Capitalism: Politics and Economics in American Thought* (Baltimore: Johns Hopkins University Press, 2006); William Martin, *With God on Our Side: The Rise of the Religious Right in America* (New York: Broadway, 1996).

2. *Engel v. Vitale*, 370 U.S. 421 (1962); *Abington Township v. Schempp* and *Murray v. Curlett*, 374 U. S. 203 (1963); *Lemon v. Kurtzman* and *Earley v. Dicenso*, 403 U. S. 602 (1971); *Stone v. Graham*, 449 U. S. 39 (1980); *Santa Fe Independent School District v. Doe*, 120 S. Ct. 2266 (2000).

3. Robert S. Alley, *Without a Prayer: Religious Expression in Public Schools* (Amherst, NY: Prometheus, 1996), pp. 46–47, 73, 210–23; Kenneth M. Dolbeare and Phillip E. Hammond, *The School Prayer Decisions: From Court Policy to Local Practice* (Chicago: University of Chicago Press, 1971); Shahrzad Habibi, "God, School, and the Fourth R: Religion Policy in Texas Public Schools," unpublished undergraduate Government Department honors thesis, 2005.

4. Martin, *With God on Our Side*, pp. 117–43.

5. *Roe v. Wade*, 410 U.S. 113 (1973).

6. "A Special Report," *Christian Harvest Times*, June, 1980, p. 1, quoted in Martin, *With God on Our Side*, p. 196.

7. Seymour Martin Lipset and Earl Raab, *The Politics of Unreason: Right-Wing Extremism in America, 1790–1970* (New York: Harper and Row, 1970), p. 466.

8. Ibid., p. 444.

9. Ibid., pp. 8, 10, 13, 15.

10. Martin, *With God on Our Side*, pp. 214–16.

11. Ibid., p. 217.

12. Langdon Gilkey, *Creationism on Trial: Evolution and God at Little Rock* (Charlottesville: University of Virginia Press), pp. 260, 268.

13. Stephen Jay Gould, *Hen's Teeth and Horse's Toes: Further Reflections in Natural History* (New York: W. W. Norton, 1983), pp. 108–109.

14. Ibid., p. 118.

15. Stephen Jay Gould, *The Mismeasure of Man*, 2nd ed. (New York: W. W. Norton, 1996), p. 23.

16. Gould, *Hen's Teeth,* p. 281.

17. Stephen Jay Gould, *An Urchin in the Storm: Essays about Books and Ideas* (New York: W. W. Norton, 1987), p. 245.

18. Stephen Jay Gould, *I Have Landed: The End of a Beginning in Natural History* (New York: Three Rivers Press, 2003), p. 215.

19. Stephen Jay Gould, *The Structure of Evolutionary Theory* (Cambridge, MA: Harvard University Press), pp. 101, 989.

20. *McLean v. Arkansas*, 529 F. Supp. 1255, 1258–1264 (ED A. K. 1982); Gould's testimony from the unofficial Stephen Jay Gould website: http://www.stephenjaygould.org.

21. *Lemon v. Kurtzman*, 403 U.S. 602 (1971).

22. Gilkey, *Creationism on Trial*, p. 275.

23. John R. Zaller, *The Nature and Origin of Mass Opinion* (Cambridge: Cambridge University Press, 1992), pp. 18, 54, 76, 93; R. Michael Alvarez and John Brehm, *Hard Choices, Easy Answers: Values, Information, and American Public Opinion* (Princeton, NJ: Princeton University Press, 2002), pp. 2, 4.

24. All survey results from http://pollingreport.com/, accessed June 14, 2007.

25. Kenneth R. Miller, "Expert Statement at the Dover, Pennsylvania, Trial," pp. 23–44, Ernan McMullen, "Plantinga's Defense of Special Creation," pp. 243–73, and Michael W. Tkacz, "Thomas Aquinas vs. the Intelligent Designers," pp. 275–82, in Robert M. Baird and Stuart E. Rosenbaum, eds., *Intelligent Design: Science or Religion?* (Amherst, NY: Prometheus Books, 2007); Gilkey, *Creationism on Trial*, pp. xxv, 11, 83; Roy Clouser, "Is Theism Compatible with Evolution?" in Robert T. Pennock, ed., *Intelligent Design Creationism and Its Critics: Philosophical, Theological, and Scientific Perspectives* (Cambridge, MA: MIT Press, 2001), pp. 513–36; Gertrude Himmelfarb, *Darwin and the Darwinian Revolution* (Chicago: Ivan R. Dee, 1996), pp. 266–67, 394–98.

26. On the general secularism of scientists: Richard Dawkins, *The God Delusion* (Boston: Houghton Mifflin, 2006), pp. 100, 102; on the willingness of some scientists to embrace religion: Howard J. Van Till, "When Faith and Reason Cooperate," in Pennock, *Intelligent Design Creationism,* pp. 147–63; Michael Ruse, "Methodological Naturalism Under Attack," in Pennock, "Intelligent Design Creationism," pp. 364–85; Gilkey, *Creationism on Trial*, pp. xxv, 11, 83.

27. Duane T. Gish, *Evolution? The Fossils Say No!* 3rd. ed. (San Diego: Creation-Life Publishers, 1979); John C. Whitcomb and Henry M. Morris, *The Genesis Flood: The Biblical Record and Its Scientific Implications* (Phillipsburg, NJ: Presbyterian and Reformed Publishing Company, 1961).

28. Michael Shermer, *Why Darwin Matters: The Case Against Intelligent Design* (New York: Henry Holt, 2006); Barbara Forrest and Paul R. Gross, *Creationism's Trojan Horse: The Wedge of Intelligent Design* (Oxford: Oxford University Press, 2004); Massimo Pugliucci, *Denying Evolution: Creationism, Scientism, and the Nature of Science* (Sunderland, MA: Sinauer Associates, 2002); Douglas J. Futyma, *Science On Trial: The Case for Evolution* (Sunderland, MA: Sinauer Associates, 1982); Philip Kitcher, *Abusing Science: The Case Against Creationism* (Cambridge, MA: MIT Press, 1982); Thomas G. Gregg, Gary R. Janssen, and J. K. Bhattacharjee, "A Teaching Guide to Evolution," National Science Teachers Association, *The Science Teacher*, 2003; Paul R. Gross, "Intelligent Design and That Vast Right-Wing Conspiracy, *Science Insights* 7, no. 4 (2002); Eugenie C. Scott, "The 'Science and Religion Movement'; An Opportunity for Improved Public Understanding of Science?" in Paul Kurtz, ed., *Science and Religion: Are They Compatible?* (Amherst, NY: Prometheus, 2003), pp. 111–25.

29. Jonathon Wells, "Common Ancestry on Trial," in William. A. Dembski, ed., *Darwin's Nemesis: Phillip Johnson and the Intelligent Design Movement* (Leicester: Inter-Varsity Press, 2006), p. 171; Phillip E. Johnson, *Darwin On Trial*, 2nd ed. (Downers Grove, IL: Inter-Varsity Press, 1993), pp. 18, 19, 27; Kenneth Poppe, *Reclaiming Science From Darwinism: A Clear Understanding of Creation, Evolution, and Intelligent Design* (Eugene, OR: Harvest House, 2006), p. 196.

30. James Barham, "Why I Am Not a Darwinist," in William A. Dembski, ed., *Uncommon Dissent: Intellectuals Who Find Darwinism Unconvincing* (Wilmington, DE: Intercollegiate Studies Institute, 2004), p. 177.

31. Ann Coulter, *Godless: The Church of Liberalism* (New York: Three Rivers Press, 2007), p. 232.

32. Stephen Jay Gould, "Return of the Hopeful Monster," in *The Panda's Thumb: More Reflections in Natural History* (New York: W. W. Norton, 1980), pp. 186–93.

33. Poppe, *Reclaiming Science*, pp. 261–71.

34. Jerry A. Coyne and H. Allen Orr, *Speciation* (Sunderland, MA: Sinauer Associates, 2004), pp. 87–91.

35. Ibid., pp. 88, 90.

36. This discussion is based on information in Coyne and Orr, *Speciation*, pp. 159–62, and Matthew J. Brauer and Daniel R. Brumbaugh, "Biology Remystified: The Scientific Claims of the New Creationists," in Pennock, *Intelligent Design Creationism*, p. 296.

37. Michael Denton, *Evolution: A Theory in Crisis* (Chevy Chase, MD: Adler and Adler, 1986), p. 117.

38. Ibid., p. 174.

39. Stephen Jay Gould, *Dinosaur in a Haystack: Reflections in Natural History* (New York: Harmony Books, 1995), p. 370.

40. Ibid., p. 369.

41. Stephen C. Meyer, "The Origin of Biological Information and the Higher Taxonomic Categories," in Dembski, *Darwin's Nemesis*, p. 175.

42. Richard Dawkins, *The Blind Watchmaker: Why the Evidence of Evolution Reveals a Universe Without Design*, 2nd ed. (New York: W. W. Norton, 1996), p. 307.

43. Kitcher, *Abusing Science*, p. 105.

44. Johnson, *Darwin On Trial*, p. 81; Timothy Standish, "Cutting Both Ways: The Challenge Posed by Intelligent Design to Traditional Christian Education," in Dembski, *Darwin's Nemesis*, p. 129; Stephen Meyer, "The Origin of Biological Information and the Higher Taxonomic Categories," in Dembski, *Darwin's Nemesis*, p. 183; Michael J. Behe, *Darwin's Black Box: The Biochemical Challenge to Evolution* (New York: Simon and Schuster, 1996), p. 95; Cornelius G. Hunter, *Darwin's God: Evolution and the Problem of Evil* (Grand Rapids, MI: Brazos Press, 2001), p. 59.

45. Arthur Koestler, *Janus: A Summing Up* (New York: Random House, 1978), pp. 165–228; Arthur Koestler, *The Case of the Midwife Toad* (New York: Random House, 1973).

46. Stephen Jay Gould, *The Structure of Evolutionary Theory* (Cambridge, MA: Harvard University Press, 2002), p. 145.

47. Michael Ruse, *Darwin and Design: Does Evolution Have a Purpose*? (Cambridge, MA: Harvard University Press, 2003), pp. 227–29; Richard Dawkins, *The Extended Phenotype: The Long Reach of the Gene*, 2nd ed. (Oxford: Oxford University Press, 1999), pp. 34, 137, 204, 209; Ernst Mayr, *Evolution and the Diversity of Life* (Cambridge, MA: Harvard University Press, 1976), pp. 40–41, 47, 95, 319; Sean B. Carroll, Jennifer K. Grenier, and Scott D. Weatherbee, *From DNA to Diversity: Molecular Genetics and the Evolution of Animal Design*, 2nd ed. (Malden, MA: Blackwell, 2005), pp. 214, 230–35.

48. Gould, *Structure*, p. 342.

49. Gould, "Return of the Hopeful Monster," p. 192.

50. Johnson, *Darwin on Trial*, p. 61.

51. Coulter, *Godless*, p. 225.

52. Poppe, *Reclaiming Science*, pp. 229, 234, 279.

53. Gould, *Panda's Thumb*, p. 20.

54. Ibid., pp. 20–21.

55. Ibid., p. 24.

56. Ibid., p. 26.

57. William A. Dembski, *Intelligent Design: The Bridge between Science and Theology* (Downers Grove, IL: InterVarsity Press, 1999), p. 261.

58. Paul A. Nelson, "The Rose of Theology in Current Evolutionary Reasoning," in Pennock, *Intelligent Design Creationism*, p. 680.

59. Stephen Jay Gould, *An Urchin in the Storm: Essays about Books and Ideas* (New York: W. W. Norton, 1987), p. 103.

60. Stephen Jay Gould and Niles Eldredge, "Punctuated Equilibrium Comes of Age," *Nature* 366 (November 1993): 223.

61. Gould, *Structure*, p. 989.

62. Gould, *Urchin in the Storm*, p. 246.

63. Stephen Jay Gould, *Rocks of Ages: Science and Religion in the Fullness of Life* (New York: Random House, 1999), p. 3.

64. Ibid., pp. 4–5.

65. Ibid, p. 6.

66. Ibid., pp. 148–149.

67. Ibid., p. 149.

68. Richard Dawkins, *A Devil's Chaplain: Reflections on Hope, Lies, Science, and Love* (Boston: Houghton Mifflin, 2003), p. 150.

69. Richard Dawkins, *The God Delusion* (Boston: Houghton Mifflin, 2006), p. 57.

70. Meyer quoted in Lee Strobel, *The Case for a Creator: A Journalist Investigates Scientific Evidence That Points Toward God* (Grand Rapids, MI: Zondervan, 2004), p. 76.

71. Dawkins, *God Delusion*, p. 57.

72. Edward O. Wilson, *Sociobiology: The New Synthesis*, 2nd ed. (Cambridge, MA: Harvard University Press, 1975, 2000), pp. 224–25; Edward O. Wilson, *On Human Nature*, 2nd ed. (Cambridge, MA: Harvard University Press, 2004), p. 116.

73. Gould, *Urchin in the Storm*, pp. 115–19; Richard Levins and Richard Lewontin, *The Dialectical Biologist* (Cambridge, MA: Harvard University Press, 1985), pp. 257–58.

74. Benjamin I. Page, *Choices and Echoes in Presidential Elections: Rational Man and Electoral Democracy* (Chicago: University of Chicago Press, 1978), pp. 152, 169, 179.

75. Thomas Hobbes, *Leviathan: Or the Matter, Forme and Power of a Commonwealth Ecclesiastical and Civil* (New York: Collier Books, 1962).

76. Gould, *Rocks of Ages*, p. 208.

77. Ibid., pp. 208, 209.

78. Shermer, *Why Darwin Matters*, pp. 116–25, 138; Barry A. Palevitz, "Science Versus Religion: A Conversation with My Students," in Kurtz, *Science and Religion*, p. 177; Mary Midgley, *Evolution as a Religion: Strange Hopes and Stranger Fears* (London: Methuen, 1985), p. 13; Gilkey, *Creationism On Trial*, p. 109; Van Till, "Faith and Reason," p. 148; Matthew Chapman, *40 Days and 40 Nights: Darwin, Intelligent Design, God, Oxycontin and Other Oddities on Trial in Pennsylvania* (New York: HarperCollins, 2007), p. 106.

79. The court case is *Tammy Kitzmiller et al. v. Dover Area School District et al.*, 400 F. Supp. 2d 707 (M. D. Pa. 2005); my generalizations are culled from many print and electronic news sources; see also Chapman, *40 Days and 40 Nights*, passim.

80. Dawkins, *Devil's Chaplain*, p. 189.

Chapter 7

A New Theory?

Any theorist of widespread influence, whether explicitly political or not—Freud, Marx, Nietzsche, Einstein, Darwin—can be accused of having put ideas into circulation that, when they came to fruition generations later, contributed to some sort of human misbehavior. Because the world never lacks for evil actions, there is always a horrifying example ready made to be cast as a logical consequence of whatever theory the observer abhors. And thus, ever since Darwin wrote, some social observers have been blaming his theory for reprehensible political trends.

Probably because arch-conservative Herbert Spencer was the first to jump on the social-Darwinist bandwagon in the nineteenth century, Western intellectuals have tended to associate the name of Charles Darwin with conservative politics. As Francis Graham Wilson summarized the dominant chain of reasoning in 1949, Darwin had convinced the world that "nature, if left alone, works for the benefit of the human race," and this concept had been connected by industrial propagandists to the idea of the "principle of the protection of property."[1] In 1961, Thomas Cochran and William Miller asserted that the political doctrine arising from the English scientist's theory, social Darwinism, had "'scientifically' justified ceaseless exploitation."[2] In 1985 Mary Midgley argued that Darwin had been misunderstood within and without biology, so that a distortion of his scientific theory to the effect that "life has been scientifically proved to be essentially competitive, in some sense which exposes all social feeling as somehow mere humbug and illusion," with the consequence that "our tradition now lacks the natural, obvious concepts by which most of the human race would denounce unbridled human predation."[3] In 1998 Marilynne Robinson informed us that "we find the recrudescence of primitive economics occurring alongside a new prominence of Darwinism. We find them separately and together encouraging faith in the value of self-interest and raw competition."[4]

But, as Gertrude Himmelfarb pointed out in her survey of the consequences of Darwinian thinking in 1959, leftist politics have also been blamed on natural selection theory.[5] And since Himmelfarb wrote, plenty of conservatives have attributed the two greatest political demons of the twentieth century to the ideas of the English naturalist. As Ben Stein informed the *New York Times* in 2007, he believed that the theory of evolution leads to racism and ultimately genocide, and he wished the new anti-Darwinist documentary he was narrating could have been titled "From Darwin to Hitler."[6] The same year, Ann Coulter, the embodiment of unrestrained conservative expression, explained, "Liberals get upset with people like Hitler and Stalin who apply Darwinism in ways they don't like, but they can't quarrel with the underlying philosophy. Certain patterns of behavior and beliefs flow naturally from the idea that man is an accident with no greater moral significance than a stalk of corn."[7]

Nevertheless, Darwinism does not have a bad name among all political thinkers. Especially on the left, there have recently been attempts to base a philosophy of social action on Darwinian values. In *A Darwinian Left?* for example, philosopher Peter Singer recounted that progressives had rejected evolutionary theory because "it dashed the left's Great Dream: The Perfectibility of Man."[8] But, he argued, the repudiation was too hasty. There is no progressive political theory based on Darwinian premises, but there should be. Although Singer did not supply the tenets of this theory, he sketched an outline of what it should do. Primarily, "We need to think about how to set up the conditions in which cooperation thrives," and "A Darwinian left, understanding the prerequisites for mutual cooperation as well as its benefits, would strive to avoid economic conditions that create outcasts."[9]

But the recommendation that political theorists come up with a new left-wing philosophy based on Darwinist principles has not stimulated a rethinking. Professional philosophers have so far been paralyzed at the prospect of constructing another progressive philosophy to take the place of the variants of Marxism, which ran out of gas in the late twentieth century. But if political philosophers have not been able to unite social and biological thought, perhaps a fresh start could be made by approaching the problem from the other side of the coin, so to speak. If political theorists have not been able to incorporate biology into politics, perhaps a scientist could incorporate politics into biology. In a sense, this is what some conservative theorists have been doing ever since Herbert Spencer invented the phrase "survival of the fittest." The challenge would be to derive the political values of equality, cooperation, compassion, and respect for governmental economic activism from a view of nature from which Spencer derived the opposite values.

Stephen Jay Gould has been part of the intellectual conversation that portrays Darwinism as best interpreted as a conservative bulwark. In some of his essays he tried to soften the received notion of Darwinian competition. He pointed out, for example, that Darwin did not, in every instance, think of competition as occurring between individual organisms. He quoted two sentences from *On the Origin of Species* in which Darwin clarified the way he used the metaphor of "struggle":

> Two canine animals, in a time of dearth, may be truly said to struggle with each other which shall get food and live. But a plant on the edge of a desert is said to struggle for life against the drought.[10]

Gould's quotation of Darwin in this instance was part of an essay urging a second look at the Russian evolutionist, Peter Kropotkin, who argued "in his cardinal premise, that the struggle for existence usually leads to mutual aid rather than combat as the chief criterion of evolutionary success."[11] Kropotkin recognized that the life of organisms did feature the conflict of individuals against each other, but he stressed that Darwin's second type of conflict—of organisms against a harsh physical environment, rather than of members of the same species against each other—was probably more important in the history of life. And that type of struggle mandated not individual competition but cooperation—mutual aid. Gould made his own position clear: "I would hold that Kropotkin's basic argument is correct."[12]

Nevertheless, Gould also recognized—indeed, he emphasized—that the basic derivation of politically conservative conclusions from Darwin's framework of ideas was historically accurate. Several times in his popular essays he repeated the point that, just before Darwin achieved his breakthrough intellectual insight as to the conceptual essence of natural selection, he had been reading economist Thomas Malthus's gloomy essay on the tendency of human populations to outrun the available food supply. Moreover, Darwin had also become familiar with Adam Smith's great thesis in *The Wealth of Nations* that competition among producers drives prices down and supply up, thus deriving public benefit from private struggle. Gould was unambiguous in characterizing the result. In the prologue to his first book he informed his readers that Darwinism was "the economy of Adam Smith transferred to nature," and he repeated the point numerous times in subsequent books.[13] Despite his qualifications and interest in the exceptions, then, Gould substantially agreed with the tendency of modern thought to ascribe politically conservative tendencies to the dominant biological theory.

Gould never announced his intention to revise Darwinism for political reasons. Indeed, as I have emphasized throughout this book, he did not separate political philosophy from scientific theory. He saw the two types of thought as linked facets of the same intellectual edifice. Perhaps his clearest statement describing his own combination of motives occurs, not in any of his formal writings, but in his testimony in the 1981 trial of the Arkansas statute mandating the teaching of "scientific creationism" in public schools.[14] The lawyer for the state, on cross-examination, asked him, "Do you have any political motivation in opposition to creation science?" Gould answered, "As Aristotle said, man is a political animal. I think everything one does is partly in the context of one's larger views."[15] In his final years his ambition was clearly to create an evolutionary theory that would modify and reform Darwinism in line with his larger political views, while retaining the respect of other scientists.

A Grand Theory?

Gould's final book, *The Structure of Evolutionary Theory*, is a combined history of evolutionary biology and an only partially completed system of ideas to reform it. I have discussed various aspects of Gould's critique of mainstream Darwinism, and his arguments for alternatives, in separate chapters of this book. Here, I will draw together his ideas as he did at much greater length in *Structure*, but, in addition, I will remark on the political implications of the changes he advocated—an activity he avoided in that book.

At the level of ontology, Gould emphasized the pitfalls of reductionism, especially the gene's-eye-view approach to biology.[16] A better way to study evolution is by adopting the outlook of *emergence*, the stance that, in Ernst Mayr's words, is based on the premise that "when two entities are combined at a higher level of integration, not all the properties of the new entity are necessarily a logical or predictable consequence of the properties of the components."[17] At every level of the hierarchy of biological reality, wrote Gould, at the levels of "genes, cells, organisms, demes, species, and clades," biologists must be prepared to recognize the "complex nonadditivity" of each level, and to analyze the evolutionary forces working on that level, independent of all the others.[18] Cells have properties that cannot be understood as deriving from the properties of genes; organisms have properties that cannot be understood by even a complete knowledge of cells; demes (local populations genetically distinct from the rest of the members of the species) have characteristics that are not deducible from organisms; and so on.

The implication of the rejection of reductionism and the embrace of emergence would be the realization that traditional Darwinist microevolution, while a fine theory with which to explain small changes within species, is inadequate to help us paint the broad patterns in the history of life. For that we need a theory of macroevolution, which would help us to explain the emergence and extinction of species, and even of levels above species. The orthodox Darwinist approach, which posits all patterns in evolution as just an extrapolation of tiny changes in the genomes of individuals, must be relegated to the analysis of small variations within species—oscillations in the beak size of finches in tandem with rainfall cycles, for example.[19] But a macroevolutionary theory needs new concepts to deal with emergent changes—concepts such as punctuated equilibria. As Gould and Eldredge developed the theory, the punctuations visible in the fossil record marked the most important events in the history of life: speciations. One set of (gradualist) concepts are very useful for investigating microevolution, but to investigate macroevolution requires another set of (nongradualist) concepts. Especially, it requires a new theory of speciation. Eldredge summarized the idea:

> Speciation involves more than simple, linear adaptive transformation of genes, behaviors, physiologies and physiognomies. It is, quintessentially, a sundering of one coherent, integrated reproductive community into two . . . discrete daughter reproductive communities.[20]

This new way of looking at the history of life would thus be a hierarchical theory, requiring different levels of explanation for handling changes at different levels of life. Of those different levels of explanation, biology has only made partial progress in constructing a theory of emergence at the species level.[21] What the theories might be that would apply to higher levels —why did the dinosaurs die while the mammals survived?—is a project for the future to produce.

At the level of methodology, Gould did not urge biologists to stop using adaptationism, but to realize that structural constraints were a very important part of the development process of life, and to seek to incorporate formalist criteria into the theory. He sought, he wrote, "a system of knowledge that requires a structure of explanation based as much on how organisms *can* be built, as on how they *do* work."[22] He urged his colleagues to expand their knowledge of the two aspects of "constraint"—"the negative concept of restriction, and the positive sense of channeling."[23] He suggested that the exploration of formalist constraints had long been stymied by the inability of biologists to identify the mechanisms of constraint. But now, he argued, the discovery of "deep homology," or regulator genes that seemed to be common to all multicelled animals since the Cambrian explosion more than 500 million years ago, held out the prospect of specifying the lines of development that restricted and channeled evolution.[24] In other words, startlingly similar developments in organisms quite widely separated on the developmental bush—the appearance of similarly designed eyes in mammals and octopi, for example—could now be understood not as comparable adaptations in different kinds of animals to comparable evolutionary problems, but as identical responses to a fundamental genetic constraint. These discoveries had caused a "general shift in viewpoint—from a preference for atomistic adaptationism . . . to a recognition that homologous developmental pathways . . . strongly shape current possibilities 'from the inside'."[25] Therefore, evolutionary biologists should now recognize that "these internal constraints can surely claim equal weight with natural selection in any full account of the causes of any particular evolutionary change."[26]

This summary should establish the fact that despite its length of 1,433 pages, Gould's final book did not offer a new theory of evolutionary change. Instead, it sketched out the desirable qualities that in his judgment a new theory should have. But a call for a theory is not a theory. Gould's legacy for evolutionary biology, then, was a series of good starts, shrewd critiques, memorable phrases, and half-baked ideas, rather than a comprehensive alternative to mainstream Darwinism. If it was not a fully formed scientific theory, it could not be fully formed political theory. His political legacy, then, must be much like his scientific legacy, a set of ideas that cohere more in tone and attitude than in conceptual completeness.

If something approaching a Gouldian theory of evolution does emerge, its political implications are fairly easy to anticipate. It will be based on fundamental skepticism of the reductionist ontology, and inclined to see organisms as wholes. Similarly, it will adopt adaptationist approaches to explanation with considerable reluctance, after having exhausted other approaches. In particular, it will be extremely skeptical of reductionist and adaptationist explanations of human behavior. It will posit a different set of mechanisms to explain evolutionary change at every level in the hierarchy. Especially, it will sever explanations of change at the genetic level from explanations at the species level. It will view survival in the pattern of life as being fundamentally contingent, interpreting the concept of "fit" as essentially the same as "lottery winner." It will assume that human behavior is largely a function of culture, and be, if not absolutely rejecting, then a tough sell for any theory of explanation based on evolutionary concepts. It will, however, agree entirely with orthodox Darwinism that all explanations should be naturalistic and based on operational concepts.

Two Good Ideas

An unkind appraisal of Gould's career would conclude that he was primarily a popularizer rather than a scientific theorist, and that his greatest contributions to evolutionary biology were his rhetorical publicizing of ideas spawned by others. Gould did not invent the concept of punctuated equilibria; rather, he made it vivid and powerful to audiences within and without biology. He did not originate the critique of the adaptationist program, functioning more as an agitator who made the profession take the attack seriously. Since Gould failed, in his final book, to provide a complete theoretical alternative to mainstream Darwinism, he might be characterized as a scientific nonentity with a genius for public relations.

But such a summary would be harshly incomplete. If Gould does not leave behind an alternative to Darwinian orthodoxy, he does bequeath a couple of related concepts that have become part of the fabric of biological thought. My intuition is that in the future, when someone with Gould's gift for historical empathy writes a history of biology in the last third of the twentieth century, he or she will recognize the concept of the spandrel, with its related concept of the exaptation, as two of the key ideas that helped to move the science forward.

The origin problem continues to be a weakness of evolutionary theory. Granted that mutations create opportunities for the environment to cull options, and thus allow organisms to adapt to conditions, what calls forth the needed mutations? The standard orthodox Darwinian reply, that given a big enough population and enough time, the normal random mutation rate will eventually throw out the necessary allele, has always satisfied some, but not all biologists. Those who have not been happy with waiting for the mutated Godot have sought various mechanisms that might create an "evolutionary director" so as to satisfy the needs of a "projective evolution."[27] But the various alternative evolutionary mechanisms—and there have been many, over the decades—have never found enough support within evolutionary biology to rid it of the origin problem.

Spandrels and exaptations, however, partially solve the origin problem without having to rely on the unorthodox concept of an evolutionary director within organisms. The spandrel is an evolutionary hitchhiker, a nonfunctional by-product that becomes functional in a new environment. Gould uses a homely example to get across the idea: "As an unintended consequence of their thinness, American dimes (and no other indigenous coins) also happen to fit snugly into the operative groove on the head of a standard screw."[28] Dimes were not designed to be useful as anything but money. The "thinness" of that coin is purely the by-product of its design as currency and the particular value of silver. But the unintended consequence of their design is that dimes are constantly put in use as screwdrivers when the actual tool is not immediately at hand. When employed in that manner, the dime becomes an exaptation, a piece of morphology or behavior created for one purpose and pressed into service for another. Thus, "American dimes are . . . adaptations as money and exaptations as screwdrivers."[29]

Not all exaptations arise from the sudden usefulness of a previously useless by-product. The term also applies to features evolved for one purpose that, under changing environmental forces, become useful for a new purpose. Gould liked to use the example of feathers. "The ancestors of birds were surely flightless and feathers did not arise all at once and fully formed. How could natural selection build an adaptation through several intermediate stages in ancestors that had no use for it?"[30] It is the sort of question that creationists typically float: Surely half a feather would be useless to a creature trying to fly, and how could such a complicated structure have arisen all at once? Gould pointed out, in several of his popular essays, that evolutionary biologists had answered this question. Feathers evolved in rudimentary form to keep dinosaurs warm. They then served as exaptations, being modified in structure, as dinosaurs evolved into birds. Thus, "feathers, evolved primarily for insulation, were soon exploited for another purpose in flight."[31] For much of his career he continued to comment on new fossil discoveries of feathered dinosaurs, and new theoretical work expanding and refining the original idea of feathers as an exaptation.[32]

The concepts of spandrels and exaptations solve part of the origin problem by rendering it moot. By avoiding the need for a creative mutation to fashion a whole new organ or behavior, by relying on presumably less-drastic mutations to develop an extant organ or behavior, they help to answer the problem of the missing intermediaries without appealing to the need for massive mutational changes in a short time span. Gould did not invent the concept of the exaptation. Biologists had been referring to "preadaptation" and "functional shift" in evolution since the nineteenth century. But by adding the idea of the spandrel as a new and fecund source of exaptations, by giving the concept a new name that did not imply teleology, and by emphasizing it in his writings, Gould expanded the imagination of his fellow professionals. Today, even the arch anti-Gould, Richard Dawkins, uses the notion of "by-products" of evolution in his writing, although he refuses to actually write the word *spandrel*.[33]

A Postscript on the History and Philosophy of Science

As part of my preparation to write a book about Gould's ideas, I put myself through an intensive reading campaign in the history and philosophy of science. I learned a great deal. The main thing I learned was that scholars in this area have neglected the political implications of scientific controversies, and the scientific implications of political controversies.

I read, for example, the forty-one articles collected in the 774 pages of *The Philosophy of Science*, edited by Richard Boyd, Philip Gasper, and J. D. Trout.[34] This volume offers significant essays by the heavyweights of twentieth-century, and occasionally nineteenth-century, philosophers: Carl

Hempel, Rudolf Carnap, Karl Popper, Thomas Kuhn, Larry Laudan, Max Weber, and a variety of others. It even has a section devoted to controversies within biology. But except for the occasional essay by Marxists, or about Marx's ideas, it contains nothing that is relevant to many of the issues I have addressed in this book. The word *politics* does not even appear in the subject index, although *ideology* rates four mentions.

That compilation is a synecdoche of the general situation. Except for the Marxists, the writing about the controversies in evolutionary biology exhibits a strange disconnect from the contextual setting of those controversies. It is almost as if writers about science are autistic, idiot savants who can brilliantly analyze the technical issues of a conflict but do not even notice the human disagreements that surround and motivate them. Or, perhaps it would be more accurate to say that writers on the subject of evolutionary biology give the impression that they think that politics is a sleazy pastime, best not noticed by polite people. I do not, as a general rule, agree with most of the Marxist political positions on the issues within evolutionary biology. But at least the Marxists admit that politics exists. The rest of the writing on these issues exhibits a one-dimensionality that renders it worse than useless—it is misleading.

By saying this I am not taking a position on the controversy between "externalists," who believe that the products of science always reflect social power, and the "internalists," who believe that the weight of the evidence and cogency of arguments determine what scientists believe.[35] I am addressing a related but different issue. Whether or not scientific ontologies and methodologies are functions of political power is itself part of the substance of scientific dispute, or should be. It does no good to say that one is an "internalist," and that one can therefore ignore the conflicts swirling around one's laboratory. The implications of scientific methodologies and conclusions are not only of interest to people outside the laboratory; they are part of the thinking of the scientists themselves. A historian should address these issues as part of the reconstruction of the scientific process. A philosopher should address these issues as part of the analysis of scientific procedure.

It was one of Gould's strengths, I think, that he was at all times fully aware of the political context of his own arguments, as well as those of his opponents. Part of the reason for his spectacular success as an essayist may have been that he did not present the scientific process as something bloodless and timeless, as a sterile abstraction of method. The fact that he argued in favor of this or against that scientific method or conclusion because of its political implications did not make his essays less compelling, but rather more compelling. He recontextualized subjects that many other scientists—and philosophers and historians—are always trying to decontextualize. But science is part of life! Science is about who gets what, about power, about moral rights and wrongs. Gould's success in presenting science this way was more interesting to nonscientists because it possessed more fidelity to the subject matter than the expurgated models of the institution that they had been fed in science classes and by public pronouncements all their lives. As an evolutionary theorist, Gould may not last. But his writing will last because by recontextualizing biological discourse he demonstrated, to scientists, to nonscientists, and even to antiscientists, why it was relevant.

notes

1. Francis Graham Wilson, *The American Political Mind: A Textbook in Political Theory* (New York: McGraw-Hill, 1949), pp. 356–57.

2. Thomas Cochran and William Miller, *The Age of Enterprise: A Social History of Industrial America* (New York: Harper, 1961); quoted, without page attribution, in Philip Abbot, *Political Thought in America: Conversations and Debates* (Prospect Heights, IL: Waveland, 1991), p. 187.

3. Mary Midgley, *Evolution as a Religion: Strange Hopes and Stranger Fears* (London: Methuen, 1985), pp. 6, 144.

4. Marilynne Robinson, *The Death of Adam: Essays on Modern Thought* (Boston: Houghton Mifflin, 1998), p. 29.

5. Gertrude Himmelfarb, *Darwin and the Darwinian Revolution* (Chicago: Ivan R. Dee, 1959, 1996), pp. 421–23, 431.

6. Stein, quoted in Cornelia Dean, "Scientists Feel Miscast in Film on Life's Origin," *New York Times*, September 27, 2007, p. A1.

7. Ann Coulter, *Godless: The Church of Liberalism* (New York: Three Rivers Press, 2007), p. 274.

8. Peter Singer, *A Darwinian Left? Politics, Evolution, and Cooperation* (New Haven, CT: Yale University, 2000), p. 24.

9. Ibid., pp. 52, 53.

10. Stephen Jay Gould, *Bully for Brontosaurus: Reflections in Natural History* (New York: W. W. Norton, 1991), p. 327.

11. Ibid., p. 331.

12. Ibid., p. 338.

13. Stephen Jay Gould, *Ever Since Darwin: Reflections in Natural History* (New York: W. W. Norton, 1977), p. 12; *The Panda's Thumb: More Reflections in Natural History* (New York: W. W. Norton, 1980), p, 66; *Dinosaur in a Haystack: Reflections in Natural History* (New York: Harmony, 1995), p. 329; *The Structure of Evolutionary Theory* (Cambridge, MA: Harvard University Press, 2002), pp. 121–25.

14. *McLean v. Arkansas*, 529 F. Supp. 1255, 1258–64 (ED A. K. 1982).

15. Transcript of Gould's testimony in *McLean v. Arkansas* from the unofficial Stephen Jay Gould website: www.stephenjaygould.org, pp. 64–65.

16. Gould, *Structure*, pp. 614–19.

17. Ernst Mayr, *Toward a New Philosophy of Biology: Observations of an Evolutionist* (Cambridge, MA: Harvard University Press, 1988), p. 34.

18. Gould, *Structure*, pp. 681–82, 628.

19. Jonathan Weiner, *The Beak of the Finch: A Story of Evolution in Our Time* (New York: Random House, 1995), pp. 193, 205, 271.

20. Niles Eldredge, *Time Frames: The Evolution of Punctuated Equilibria* (Princeton, NJ: Princeton University Press, 1985), p. 97.

21. See the summary of current research in Jerry A. Coyne and H. Allen Orr, *Speciation* (Sunderland, MA: Sinauer Associates, 2004).

22. Gould, *Structure*, p. 312.

23. Ibid., p. 323.

24. Ibid., p. 1056.

25. Ibid., p. 1123.

26. Ibid., p. 1174.

27. John H. Campbell, "An Organizational Interpretation of Evolution," in David J. Depew and Bruce H. Weber, eds., *Evolution at a Crossroads: The New Biology and the New Philosophy of Science* (Cambridge, MA: MIT Press, 1985), pp. 143, 144, 146, 148.

28. Gould, *Structure*, p. 1277.

29. Ibid., p. 1278.

30. Gould, *Panda's Thumb*, p. 276.

31. Ibid.

32. Stephen Jay Gould, *I Have Landed: The End of a Beginning in Natural History* (New York: Three Rivers Press, 2003), pp. 321–32.

33. Richard Dawkins, *The God Delusion* (Boston: Houghton Mifflin, 2006), p. 172.

34. Richard Boyd, Philip Gasper, and J. D. Trout, ed., *The Philosophy of Science* (Cambridge, MA: MIT Press, 1991).

35. David L. Hull, *Science as a Process: An Evolutionary Account of the Social and Conceptual Development of Science* (Chicago: University of Chicago Press, 1988), p. 1.

Glossary

adaptationist program: The dominant research strategy in evolutionary biology, in which researchers ask themselves what might be the adaptive value of a bodily form or behavior; based on the assumption that all morphology and behavior are adaptive. (chapter 4)

allele: Any one of possible different forms of a gene that can occupy the same place (locus) on a chromosome. (chapter 7)

allometry: The study of the relative growth of one or some parts of an organism in comparison with other parts or the whole. (chapter 4)

contingency: Happenings in history governed by luck and probability, such that they are not predictable in advance, even with perfect knowledge of conditions. (chapter 3)

convergence: The repeated tendency in the animal world for different lineages to arrive at very similar solutions to evolutionary problems, as when both squids and mammals independently evolved eyes with similar structure. (chapter 3)

craniology: The doctrine that brain size, or, more generally, head size, determines intelligence. (chapter 5)

critical elections: In one theory of American politics, an election or series of elections in which significant groups of voters either switch their party allegiance more or less permanently or cease to participate at the ballot box, in either case having important consequences for the partisan balance of power. (chapter 3)

deme: A local population largely isolated genetically from the other members of the species, featuring clearly definable genetic characteristics. (chapter 7)

emergence: The situation in which a combination of two or more substances exhibits characteristics that occur in none of the original substances individually. (chapter 7)

empiricism: In philosophy, the doctrine that all knowledge is based on and exhausted by what is known from sensory experience; as used in science, generally means the doctrine that the only appropriate form of evidence is that which can be demonstrated to the senses of other observers. (chapter 2)

epistasis: A form of genetic interaction in which one gene interferes with the expression of another gene. (chapter 6)

epistemology: The branch of philosophy that investigates the origin, structure, methods, and validity of knowledge; frequently employed for methodological criticism in science. (chapter 2)

ethology: The study of animal behavior in the animal's natural environment rather than in the laboratory. (chapter 2)

exaptation: A feature of an organism, either of morphology or behavior, that is an adaptation of a previously nonadaptive by-product (spandrel), or of a feature adapted for some other purpose; for example, feathers, which originally evolved to keep dinosaurs warm but were adapted by birds into flight aids. (chapter 4)

externalism: A philosophical doctrine holding that truth is not objective, but is instead a product of social context, making science an activity that cannot produce objective facts about reality. (chapter 2)

formalism: The approach to explaining the form and behavior of organism that emphasizes structural constraints rather than adaptive fluidity. (chapter 4)

hierarchical theory: The approach to explaining the pattern of life's history that emphasizes different types of explanations at different levels of organisms; one explanation for genes, one for organisms, one for species, and so on. (chapter 3)

holism: The methodological view that wholes are more than the sum of their parts, and that therefore reductionist science cannot produce truth. (chapter 2)

homeotic genes: Regulatory genes that control the development and expression of other genes that actually create the structures and behaviors of the organism. (chapter 6)

homology: Situation that occurs when features of different organisms that may be structurally dissimilar nevertheless owe their origins to the same structural features because of the common ancestry of the organisms; the hand of a human, the flipper of a dolphin, and the wing of a bat are homologous structures. (chapter 7)

integrative biology: One term covering the combination of biology, paleontology, chemistry, physics, and engineering that is the label used to identify modern evolutionary biology in some universities; given other labels at other schools. (chapter 3)

intelligent design: As an explanation for the arrival and forms of life, a modern variation on creationism; given the alternate label in an attempt to avoid constitutional strictures against teaching religious doctrines in public schools. (chapter 6)

internalism: A philosophical doctrine holding that truth is objective and at least approximately knowable, making science an activity that is at least partly independent of social context. (chapter 2)

Lamarckism: A theory of evolutionary change that has two main characteristics; first, that organisms can inherit acquired characteristics, and second, that organisms have some control over their own mutations, so that their offspring evolve in an adaptive direction. (chapter 6)

macroevolution: Significant evolutionary change above the species level; especially, the creation of new species. (chapter 3)

metaphor: A figure of speech in which one kind of object or idea is used to suggest a likeness or analogy to another kind of object or idea. (chapter 1)

microevolution: Significant evolutionary change within species. (chapter 3)

morphology: The study of the structure, overall form, or general appearance of an organism, as opposed to *anatomy*, which involves dissection; also, the shape or overall appearance itself. (chapter 3)

naturalistic fallacy: To philosophers, an error in reasoning in which moral lessons ("ought") are deduced from natural facts ("is"). (chapter 2)

natural selection: Darwin's proposed mechanism driving evolutionary change, proposing that the environment culls better-adapted variants within each species, which allows the survivors to propagate their genes. (chapter 3)

NOMA: "Non-overlapping magisteria"; Gould's label for his recommendation that science and religion each recognize the superiority of each other in two areas of life: religion in morals, and science in factual reality. (chapter 6)

ontology: The study of the fundamental essences of being; primary categories of existence. (chapter 2)

paleontology: Scientific discipline addressed to studying and explaining the broad patterns of life's history, using fossils as its primary source of information. (chapters 1 and 2)

phenotype: The observable features of an organism. (chapter 3)

phyletic gradualism: The slow and steady transformation, over geological time, of one morphological form into another, or one species into another; coined by Eldredge and Gould in their article on punctuated equilibria. (chapter 3)

phylum: One of the primary divisions in the Linnaean system of biological classification; less general than a kingdom, more general than a class. (chapter 3)

pleiotropy: The genetic process by which one gene affects two or more aspects of an organism that are apparently unrelated. (chapter 6)

polygenism: The doctrine that different human races originally evolved as separate species. (chapter 5)

punctuated equilibria: The theory that the pattern of evolutionary history is better conceptualized as one in which species are in morphological stasis for long periods of time, then rather quickly give birth to new species, rather than one in which evolution is conceptualized as occurring at a relatively slow, steady pace of tiny changes. (chapter 3)

realignment: From a theory about partisan behavior in American politics, a more-or-less rapid and long-lasting shift in the voting allegiances of significant groups of citizens. (chapter 3)

realism: The philosophical doctrine holding that the phenomena of the world exist independently of human thought. (chapter 2)

recapitulation: The conviction that the adults of inferior human groups must be like the children of superior groups. (chapter 5)

reductionism: The approach to scientific explanation that emphasizes two related but separate activities, the attempt to analyze phenomena both at the level of the smallest feasible entity and at the level of the most general theoretical construct. (chapter 2)

reification: An error in reasoning in which an abstract category, existing only in the human mind, is mistaken for a real empirical entity. (chapter 5)

r-squared: In statistical analysis, a measure of the goodness of fit of the data points around a regression line; specifically, the percentage of the variance around the mean accounted for by the regression equation. (chapter 5)

saltationism: The evolutionary theory based on the assumption that evolutionary change, particularly speciation, proceeds through rapid transformations, generally in a single generation; often based on a macromutation. (chapter 6)

secular humanism: A complex of nonreligious attitudes, philosophies, and policy preferences, of which Darwinism is one, which are frequently blamed by social conservatives for modern social ills. (chapter 6)

social Darwinism: Specifically, a philosophical movement during the nineteenth century holding to the tenet that people who are a success in society are, by analogy, like the "fit" in biology, and that, therefore, government should not attempt to regulate the economy or redistribute wealth and power; more generally, any philosophical argument reasoning from the premise that social and political practice should be consistent with evolved human nature. (chapters 1 and 2)

sociobiology: The study of behavior, usually social behavior, within the context of the adaptationist program. (chapter 4)

spandrel: A nonadaptive byproduct of an adaptive feature of an organism; the term is borrowed from architecture. (chapter 4)

species: The lowest group of organisms that is at least potentially capable of interbreeding, and which is reproductively isolated from other such groups. (chapter 3)

stigmata: Within biology, the notion that potential criminals can be recognized before they have committed any crimes, with regard to certain telltale external characteristics—unusually large jaws, for example. (chapter 5)

teleology: The philosophical doctrine that entities, and organisms in particular, should be considered with reference to their ends or functions and not simply with reference to their antecedent causes. (chapter 7)

trilobite: An extinct arthropod from the class *Trilobita*, the members of which lived during the Cambrian and Silurian geological epochs. (chapter 3)

vitalism: The biological doctrine that evolution cannot be understood solely in physical terms, because there is a force in life that (unconsciously) understands and achieves the mutational changes necessary for organisms to adapt their offspring to the environment. (chapter 6)

windowpane theory: The opinion that scientific discourse should be precise, factual, fair-minded, dispassionate, nonevaluative, and above all nonrhetorical. (chapter 1)

zygote: The cell produced by a fusion of the male sperm with the female egg at fertilization. (chapter 6)

Bibliography

Abington Township v. Schempp and *Murray v. Curlett*, 374 U. S. 203 (1963).

Alcock, John. *The Triumph of Sociobiology*. Oxford: Oxford University Press, 2001.

———. "Unpunctuated Equilibrium in the Natural History Essays of Stephen Jay Gould." *Evolution and Human Behavior* 19 (1998): 321–36.

Alford, John R., Carolyn L. Funk, and John R. Hibbing. "Are Political Orientations Genetically Transmitted?" *American Political Science Review* 99, no. 2 (May 2005): 153–68.

Allen, E., B. Beckwith, J. Beckwith, S. Chorover, D. Culver, M. Duncan, S. J. Gould, R. Hubbard, H. Inouye, A. Leeds, R. Lewontin, C. Madansky, L. Miller, R. Pyeritz, M. Rosenthal, and H. Schreier. "Against 'Sociobiology'." *New York Review of Books* 22 (1975): 43–44.

Allen, K. C., and D. E. G. Briggs, eds. *Evolution and the Fossil Record.* Washington, DC: Smithsonian Institution, 1990.

Alley, Robert S. *Without a Prayer: Religious Expression in Public Schools.* Amherst, NY: Prometheus, 1996.

Alroy, John. "Constant Extinction, Constrained Diversification, and Uncoordinated Stasis in North American Mammals." *Palaeo* 127 (1996): 293.

Alvarez, R. Michael, and John Brehm. *Hard Choices, Easy Answers: Values, Information, and American Public Opinion.* Princeton, NJ: Princeton University Press, 2002.

Ardrey, Robert. *The Territorial Imperative: A Personal Inquiry into the Animal Origins of Property and Nations.* New York: Kodansha, 1997.

Baird, Robert M., and Stuart E. Rosenbaum, eds. *Intelligent Design: Science or Religion?* Amherst, NY: Prometheus, 2007.

Barash, David. *The Whisperings Within: Evolution and the Origin of Human Nature*. New York: Penguin, 1979.

Barkow, Jerome H., Leda Cosmides, and John Tooby, eds. *The Adapted Mind: Evolutionary Psychology and the Generation of Culture.* New York: Oxford University Press, 1992.

Bastian, Veneta, Nicholas R. Burns, and Ted Nettelbeck. "Emotional Intelligence Predicts Life Skills, but Not as Well as Personality and Cognitive Abilities." *Personality and Individual Differences* 39, no. 6 (October 2004): 1135–45.

Baumgartner, Frank R., and Bryan D. Jones. *Agendas and Instability in American Politics*. Chicago: University of Chicago Press, 1993.

Bazerman, Charles. "Intertextual Self-Fashioning: Gould and Lewontin's Representation of the Literature." In Selzer, *Understanding Scientific Prose*, pp. 20–41.

Beck, Paul Allen. "A Socialization Theory of Partisan Realignment." In Richard Niemi and associates, *The Politics of Future Citizens*. San Francisco: Jossey-Bass, 1974.

Behe. Michael J. *Darwin's Black Box: The Biochemical Challenge to Evolution*. New York: Simon and Schuster, 1996.

Bergson, Henri. *Creative Evolution*. New York: Barnes and Noble, 2005.

Bleier, Ruth. *Science and Gender: A Critique of Biology and Its Theories on Women*. Oxford: Pergamon Press, 1984.

Blinkhorn, Steve. "What skullduggery?" *Nature* 296 (April 1982): 506–508.

Bohman, James. *New Philosophy of Social Science: Problems of Indeterminacy*. Cambridge, MA: MIT Press, 1991.

Bowler, Peter B. *Evolution: The History of an Idea*. Berkeley: University of California Press, 2003.

Boyd, Richard, Philip Gasper, and J. D. Trout, eds. *The Philosophy of Science*. Cambridge, MA: MIT Press, 1991.

Boyd, Robert, and Peter J. Richerson. *Culture and the Evolutionary Process*. Chicago: University of Chicago Press, 1985.

Brady, David W. *Critical Elections and Congressional Policy Making*. Stanford, CA: Stanford University Press, 1988.

Brauer, Matthew J., and Daniel R. Brumbaugh. "Biology Remystified: The Scientific Claims of the New Creationists." In Pennock, *Intelligent Design Creationism*, pp. 289–334.

Brooks, Cleanth, and Robert Penn Warren. *Understanding Poetry*. New York: Holt, Rinehart, and Winston, 1938.

Brown, Andrew. *The Darwin Wars: The Scientific Battle for the Soul of Man*. London: Simon and Schuster, 1999.

Brown, Richard Harvey. "Rhetoric and the Science of History: The Debate Between Evolutionism and Empiricism as a Conflict of Metaphors." *Quarterly Journal of Speech* 72 (1986): 149.

Burnham, Walter D. "The Changing Shape of the American Political Universe." *American Political Science Review* 59 (1965): 7–28.

———. "Constitutional Moments and Punctuated Equilibria: A Political Scientist Confronts Bruce Ackerman's 'We The People.'" *Yale Law Journal* 108 (1999): 2237–77.

———. *Critical Elections and the Mainsprings of American Politics*. New York: W. W. Norton, 1970.

———. "Party Systems and the Political Process." In William N. Chambers and Walter Dean Burnham, eds., *The American Party Systems: Stages of Political Development*. New York: Oxford University Press, 1967.

Buss, David M. *The Evolution of Desire: Strategies of Human Mating*. New York: HarperCollins, 1994.

Canguilhem, Georges. *Ideology and Rationality in the History of the Life Sciences*. Cambridge, MA: MIT Press, 1988.

Carroll, John B. "What Is Intelligence?" In Sternberg and Detterman, *What Is Intelligence?*, pp. 51–54.

Carroll, Robert L. *Patterns and Processes of Vertebrate Evolution*. Cambridge: Cambridge University Press, 1997.

Carroll, Sean B., Jennifer K. Grenier, and Scott D. Weatherbee. *Endless Forms Most Beautiful: The New Science of Evo Devo*. New York: W. W. Norton, 2005.

———. *From DNA to Diversity: Molecular Genetics and the Evolution of Animal Design*, 2nd ed. Malden, MA: Blackwell, 2005.

Carruthers, Peter and Andrew Chamberlain. "Introduction." In Carruthers and Chamberlain, eds., *Evolution and the Human Mind: Modularity, Language, and Meta-Cognition*. Cambridge: Cambridge University Press, 2000), pp. 1–12.

Chambers, Paul. *Bones of Contention: The Archaeopteryx Scandals*. London: John Murray, 2002.

Charles, J. Daryle. Review of *In Defense of History*, by Richard J. Evans. *Academic Questions* 13 (Winter 2000): 92–94.

Charney, Davida. "A Study In Rhetorical Reading: How Evolutionists Read 'The Spandrels of San Marco'." In Selzer, *Understanding Scientific Prose*, pp. 203–31, 299–399.

Charney, Evan. "Genes and Ideologies." *Perspectives on Politics* 6, no. 2 (June 2008): 299–319.

Clouser, Roy. "Is Theism Compatible with Evolution?" In Pennock, *Intelligent Design Creationism*, pp. 513–36.

Clubb, Jerome M., William H. Flanigan, and Nancy H. Zingale. *Partisan Realignment: Voters, Parties, and Government in American History*. Beverly Hills, CA: Sage, 1980.

Cooper, Colin. *Intelligence and Abilities*. London: Routledge, 1999.

Coyne, Jerry A., and H. Allen Orr. *Speciation*. Sunderland, MA: Sinauer Associates, 2004.

Darwin, Charles. *On the Origin of Species by Means of Natural Selection*, 1st edition. New York: Barnes and Noble Classics, 2004 [1859].

———. *On the Origin of Species by Means of Natural Selection*, 6th edition. New York: Macmillan, 1962 [1872].

Dawkins, Richard. *The Blind Watchmaker: Why The Evidence of Evolution Reveals a Universe Without Design*, 2nd. ed. New York: W. W. Norton, 1996.

———. *Climbing Mount Improbable*. New York: W. W. Norton, 1996.

———. *A Devil's Chaplain: Reflections on Hope, Lies, Science, and Love*. New York: Houghton Mifflin, 2003.

———. *The Extended Phenotype: The Long Reach of the Gene*. Oxford: Oxford University Press, 1982.

———. *The God Delusion*. Boston: Houghton Mifflin, 2006.

———. *The Selfish Gene*. Oxford: Oxford University Press, 1976.

———. *Unweaving the Rainbow: Science, Delusion, and the Appetite for Wonder*. Boston: Houghton, Mifflin, 1998.

Dean, Cornelia. "How Quantum Physics Can Teach Biologists About Evolution." *New York Times*, July 5, 2005, p. D2.

de Chardin, Pierre Teilhard. *Christianity and Evolution*. New York: Harcourt, Brace, Jovanovich, 1969.

Degler, Carl N. *In Search of Human Nature: The Decline and Revival of Darwinism in American Social Thought*. New York: Oxford University Press, 1991.

Dembski, William A., ed. *Darwin's Nemesis: Phillip Johnson and the Intelligent Design Movement*. Leicester, UK: Inter-Varsity Press, 2006.

———. *Intelligent Design: The Bridge Between Science and Theology*. Downer's Grove, IL: Inter-Varsity Press, 1999.

———, ed. *Uncommon Dissent: Intellectuals Who Find Darwinism Unconvincing*. Wilmington, DE: Intercollegiate Studies Institute, 2004.

Dennett, Daniel C. *Darwin's Dangerous Idea: Evolution and the Meanings of Life*. New York: Simon and Schuster, 1995.

Denton, Michael. *Evolution: A Theory in Crisis*. Chevy Chase, MD: Adler and Adler, 1986.

Desmond, Adrian, and James Moore. *Darwin: The Life of a Tormented Evolutionist*. New York: W. W. Norton, 1991.

Diamond, Jared. *Guns, Germs, and Steel: The Fates of Human Societies*. New York: W. W. Norton, 1999.

Dickens, William T., and James R. Flynn. "Black Americans Reduce the Racial IQ Gap: Evidence from Standardization Samples." *Psychological Science* 17, no. 10 (October 2006): 913–20.

———. "Commentary: Common Ground and Differences." *Psychological Science* 17, no. 10 (October 2006): 923–24.

Dobzhansky, Theodosius. *The Biological Basis of Human Freedom*. New York: Columbia University Press, 1956.

———. *Genetics and the Origin of Species*. New York: Columbia University Press, 1982 [1937].

Dolbeare, Kenneth M., and Phillip E. Hammond. *The School Prayer Decisions: From Court Policy to Local Practice*. Chicago: University of Chicago Press, 1971.

Duckworth, Angela L., and Martin E. P. Seligman. "Self-Discipline Outdoes IQ in Predicting Academic Performance of Adolescents." *Psychological Science* 16, no. 2 (December 2005): 939–44.

Dunnette, Marvin D. "Fads, Fashions, and Folderol in Psychology." *American Psychologist* 21 (April 1966): 343–52.

Edey, Maitland A., and Donald C. Johanson. *Blueprints: Solving the Mystery of Evolution*. New York: Penguin, 1989.

Eldredge, Niles. "The Allopatric Model and Phylogeny in Paleozoic Invertebrates." *Evolution* 25, no. 1 (March 1971): 156–67.

———. *The Miner's Canary: Unraveling the Mysteries of Extinction*. Princeton, NJ: Princeton University Press, 1991.

———. "Punctuated Equilibria, Rates of Change, and Large-Scale Entities in Evolutionary Systems." In Somit and Peterson, *Dynamics of Evolution*.

———. *Reinventing Darwin: The Great Evolutionary Debate*. London: Phoenix, 1995.

———. *Time Frames: The Evolution of Punctuated Equilibria*. Princeton, NJ: Princeton University Press, 1985.

———. *Why We Do It: Rethinking Sex and the Selfish Gene*. New York: W. W. Norton, 2004.

Eldredge, Niles, and Stephen Jay Gould. "Punctuated Equilibria: An Alternative to Phyletic Gradualism." In Thomas J. M Schopf, ed., *Models in Paleobiology*. San Francisco: Freeman, Cooper and Company, 1972, pp. 82–115.

Engel v. Vitale, 370 U. S. 421 (1962).

Engels, Frederick. *Dialectics of Nature*. Moscow: Language Publishing House, 1954.

Fahnestock, Jeanne. "Tactics of Evaluation in Gould and Lewontin's 'The Spandrels of San Marco'." In Selzer, *Understanding Scientific Prose*, pp. 158–79.

Fisher, R. A. *The Genetical Theory of Natural Selection*. Oxford: Oxford University Press, 1930.

Forrest, Barbara and Paul R. Gross. *Creationism's Trojan Horse: The Wedge of Intelligent Design*. Oxford, Oxford University Press, 2004.

Friedman, Naomi, Akira Miyake, Robin P. Corley, Susan E. Young, John C. DeFries, and John K. Hewitt. "Not All Executive Functions Are Related to Intelligence." *Psychological Science* 17, no. 2 (February 2006): 172–79.

Futuyma, Douglas J. *Science On Trial: The Case for Evolution*. Sunderland, MA: Sinauer Associates, 1982.

Galison, Peter, and David J. Stump, eds. *The Disunity of Science: Boundaries, Contexts, and Power*. Stanford, CA: Stanford University Press, 1996.

Gans, Carl. "Punctuated Equilibria and Political Science: A Neontological View." *Politics and the Life Sciences* 5, no. 2 (February 1987): 225.

Gaspar, Philip. "Causation and Explanation." In Boyd, Gasper, and Trout, *The Philosophy of Science*, pp. 289–92.

Gilkey, Langdon. *Creationism on Trial: Evolution and God at Little Rock*. Charlottesville: University of Virginia Press, 1998.

Gingerich, Philip D. "Punctuated Equilibria—Where Is the Evidence?" *Systematic Zoology* 33, no. 3 (September 1984): 335–38.

———. "Species in the Fossil Record: Concepts, Trends, and Transitions." *Paleobiology* 11, no. 1 (1985): 27–41.

Gish, Duane T. *Evolution? The Fossils Say No!* 3rd ed. San Diego: Creation-Life Publishers, 1979.

Godfrey-Smith, Peter. *Theory and Reality: An Introduction to the Philosophy of Science*. Chicago: University of Chicago Press, 2003.

Gould, Stephen Jay. *Bully for Brontosaurus: Reflections in Natural History*. New York: W. W. Norton, 1991.

———. "Curveball." In Steven Fraser, ed., *The Bell Curve Wars: Race, Intelligence, and the Future of America*. New York: HarperCollins, 1995.

———. *Dinosaur in a Haystack*. New York: Harmony Books, 1995.

———. *Eight Little Piggies: Reflections in Natural History*. New York: W. W. Norton, 1993.

———. *Ever Since Darwin: Reflections in Natural History*. New York: W. W. Norton, 1977.

———. *The Flamingo's Smile: Reflections in Natural History*. New York: W. W. Norton, 1985.

———. "Foreward." In Kirtley F. Mather, *The Permissive Universe*. Albuquerque: University of New Mexico Press, 1986.

———. "Fulfilling the Spandrels of World and Mind." In Selzer, *Understanding Scientific Prose*, pp. 310–36.

———. *Full House: The Spread of Excellence from Plato to Darwin*. New York: Three Rivers Press, 1996.

———. *The Hedgehog, The Fox, and the Magister's Pox*. New York: Three Rivers Press, 2003.

———. *Hen's Teeth and Horse's Toes: Further Reflections in Natural History*. New York: W. W. Norton, 1983.

———. *I Have Landed: The End of a Beginning in Natural History*. New York: Three Rivers Press, 2003.

———. "Is a New and General Theory of Evolution Emerging?" *Paleobiology* 6, no. 1 (1980): 119–30.

———. *Leonardo's Mountain of Clams and the Diet of Worms*. New York: Three Rivers Press, 1998.

———. *The Lying Stones of Marrakech: Penultimate Reflections In Natural History*. New York: Harmony Books, 2000.

———. *The Mismeasure of Man*, 1st ed. New York: W. W. Norton, 1981.

———. *The Mismeasure of Man*, 2nd ed. New York: W. W. Norton, 1996.

———. *The Panda's Thumb: More Reflections on Natural History*. New York: W. W. Norton, 1980.

———. *Rocks of Ages: Science and Religion in the Fullness of Life*. New York: Vintage, 1999.

———. *The Structure of Evolutionary Theory*. Cambridge, MA: Harvard University, 2002.

———. *Time's Arrow, Time's Cycle: Myth and Metaphor in the Discovery of Geological Time*. Cambridge, MA: Harvard University Press, 1987.

———. *Triumph and Tragedy in Mudville: A Lifelong Passion for Baseball*. New York: W. W. Norton, 2003.

———. *An Urchin in the Storm: Essays about Books and Ideas*. New York: W. W. Norton, 1987.

———. *Wonderful Life: The Burgess Shale and the Nature of History*. New York: W. W. Norton. 1989.

Gould, Stephen Jay, and Niles Eldredge. "Punctuated Equilibria: The Tempo and Mode of Evolution Reconsidered." *Paleobiology* 3 (1977): 115–51.

———. "Punctuated Equilibrium Comes of Age." *Nature* 366 (November 1993): 223–27.

Gould, Stephen Jay, and Richard C. Lewontin. "The Spandrels of San Marco and the Panglossian Paradigm: A Critique of the Adaptationist Paradigm." *Proceedings of the Royal Society of London* (1979 B): 147–64.

Gould, Stephen Jay, and Elisabeth S. Vrba. "Exaptation—A Missing Term in the Science of Form." *Paleobiology* 8, no. 1 (1982): 4–15.

Gowdy, John M. "New Controversies in Evolutionary Biology: Lessons for Economics." *Methodus* (June 1991): 86.

Gragson, Gay, and Jack Selzer. "The Reader in the Text of 'The Spandrels of San Marco'." In Selzer, *Understanding Scientific Prose*, pp. 180–202.

Gregg, Thomas G., Gary R. Janssen, and J. K. Bhattacharjee. "A Teaching Guide to Evolution." National Science Teachers Association, *The Science Teacher*, 2003.

Grene, Marjorie, "Perception, Interpretation, and the Sciences: Toward a New Philosophy of Science." In David J. Depew and Bruce H. Weber, eds., *Evolution at a Crossroads: The New Biology and the New Philosophy of Science*. Cambridge, MA: MIT Press, 1985, pp. 1–20.

Gross, Paul R. "Intelligent Design and That Vast Right-Wing Conspiracy." *Science Insights* 7, no. 4 (2002).

Habibi, Shahrzad. "God, School, and the Fourth R: Religion Policy in Texas Public Schools." Unpublished undergraduate Government Department honors thesis, University of Texas at Austin, 2005.

Hale, W. G., and J. P. Margham. *The HarperCollins Dictionary of Biology*. New York: HarperCollins, 1991.

Hart, Roderick P. *Modern Rhetorical Criticism*, 2nd. ed. Boston: Allyn and Bacon, 1997.

Hempel, Carl. *Philosophy of Natural Science*. Englewood Cliffs, NJ: Prentice-Hall, 1966.

Herrnstein, Richard. "IQ." *Atlantic Monthly*, September, 1971, pp. 43–67.

Herrnstein, Richard J., and Charles Murray. *The Bell Curve: Intelligence and Class Structure in American Life*. New York: Free Press, 1994.

Himmelfarb, Gertrude. *Darwin and the Darwinian Revolution*. Chicago: Ivan R. Dee, 1996 [1959].

Hobbes, Thomas. *Leviathan: Or the Matter, Forme and Power of a Commonwealth Ecclesiastical and Civil*. New York: Collier Books, 1962.

Hoffman, Antoni. *Arguments on Evolution: A Paleontologist's Perspective*. New York: Oxford University Press, 1989.

Howell, Donna J. "Plant-loving Bats, Bat-loving Plants." *Natural History* 85 (February 1976): 52–59.

Hull, David L. *Science as a Process: An Evolutionary Account of the Social and Conceptual Development of Science*. Chicago: University of Chicago Press, 1988.

Hunter, Cornelius G. *Darwin's God: Evolution and the Problem of Evil*. Grand Rapids, MI: Brazos Press, 2001.

Hutcheon, Pat Duffy. *Leaving the Cave: Evolutionary Naturalism in Social-Scientific Thought*. Waterloo, ON: Wilfred Laurier University, 1996.

Ingber, Donald E. "The Architecture of Life," *Scientific American* (January 1998): 48–57.

Jackson, Jeremy B. C., and Douglas H. Erwin. "What Can We Learn about Ecology and Evolution from the Fossil Record?" *Trends in Ecology and Evolution* 21, no. 6 (June 2006): 322–28.

Jefferson, Thomas. Letter to John Adams, October 28, 1813. In Adrienne Koch and William Peden, eds., *The Life and Selected Writings of Thomas Jefferson*. New York: Modern Library, 1944, pp. 632–34.

Jensen, Arthur R. "How Much Can We Boost IQ and Scholastic Achievement?" *Harvard Educational* Review 33 (1969): 1–123.

———. "Intelligence: 'Definition,' Measurement, and Future Research." In Sternberg and Detterman, *What Is Intelligence*?, pp. 109–12.

Johnson, Philip E. *Darwin On Trial*. Downers Grove, IL: InterVarsity Press, 1993.

Key, V. O., Jr. "A Theory of Critical Elections." *Journal of Politics* 17 (1955): 3–18.

Kitcher, Philip. *Abusing Science: The Case Against Creationism*. Cambridge, MA: MIT Press, 1982.

———. *The Advancement of Science: Science Without Legend, Objectivity Without Illusions*. New York: Oxford University Press, 1993.

———. "1953 and All That: A Tale of Two Sciences." In Boyd, Gasper, and Trout, *Philosophy of Science*, pp. 553–70.

———. *Vaulting Ambition: Sociobiology and the Quest for Human Nature*. Cambridge, MA: MIT Press, 1985.

Koestler, Arthur. *The Case of the Midwife Toad*. New York: Random House, 1973.

———. *Janus: A Summing Up*. New York: Random House, 1978.

Krieger, Leonard. *Time's Reasons: Philosophies of History Old and New*. Chicago: University of Chicago Press, 1989.

Kuhn, Thomas. *The Structure of Scientific Revolutions*, 3rd ed. Chicago: University of Chicago Press, 1996.

Kurtz, Paul, ed. *Science and Religion: Are They Compatible?* Amherst, NY: Prometheus, 2003.

Laidra, Kaia, Helle Pullmann, and Juri Allik. "Personality and Intelligence as Predictors of Academic Achievement: A Cross-Sectional Study from Elementary to Secondary School." *Personality and Individual Differences* 42, no. 3 (February 2007): 441–51.

Lemon v. Kurtzman and *Earley v. Dicenso*, 403 U. S. 602 (1971).

Levins, Richard, and Richard Lewontin. *The Dialectical Biologist*. Cambridge, MA: Harvard University Press, 1985.

Levinton, Jeffrey S. *Genetics, Paleontology, and Macroevolution*. Cambridge: Cambridge University Press, 2001.

Lewontin, Richard. *Biology as Ideology: The Doctrine of DNA*. Concord, ON: House of Anansi, 1991.

———. *The Triple Helix: Gene, Organism, and Environment*. Cambridge, MA: Harvard University Press, 2000.

Lewontin, Richard C., and Richard Levins. "Stephen Jay Gould: What Does It Mean to Be a Radical?" *Monthly Review* 54, no. 6 (November 2002): 1–8.

Lewontin, R. C., Steven Rose, and Leon J. Kamin. *Not in Our Genes: Biology, Ideology, and Human Nature*. New York: Pantheon Books, 1984.

Lipset, Seymour Martin, and Earl Raab. *The Politics of Unreason: Right-Wing Extremism in America, 1790–1970*. New York: Harper and Row, 1970.

Lloyd, Christopher. *The Structures of History*. Cambridge: Blackwell, 1993.

Locke, John. *Second Treatise of Government*. Indianapolis: Hackett, 1980.

Loehlin, John C. "Should We Do Research on Race Differences in Intelligence?" *Intelligence* 16 (1992): 1–4.

Lorenz, Konrad. *On Aggression*. New York: Bantam, 1966.

Loring, Brace C., and Frank B. Livingstone. "On Creeping Jensenism." In Montagu, *Race and IQ*, pp. 215–16.

Lyne, John, and Henry F. Howe. "'Punctuated Equilibria'; Rhetorical Dynamics of a Scientific Controversy." *Quarterly Journal of Speech* 72 (1986): 132–47.

Martin, William. *With God on Our Side: The Rise of the Religious Right in America.* New York: Broadway, 1996.

Marx, Karl, and Friedrich Engels. "The Communist Manifesto." In Arthur P. Mendel, ed., *Essential Works of Marxism.* New York: Bantam, 1965.

Masters, Roger D. *The Nature of Politics*. New Haven, CT: Yale University Press, 1989.

Mather, Kirtley F. *The Permissive Universe*. Albuquerque: University of New Mexico, 1986.

Mayhew, David R. *Electoral Realignments: A Critique of an American Genre*. New Haven, CT: Yale University Press, 2002.

Mayr, Ernst. "Change of Genetic Environment and Evolution." In Mayr, *Evolution and the Diversity of Life.*

———. *Evolution and the Diversity of Life*. Cambridge, MA: Harvard University Press, 1976.

———. Introduction to "Change of Environment and Speciation." In Mayr, *Evolution and the Diversity of Life.*

———. *Toward a New Philosophy of Biology: Observations of an Evolutionist.* Cambridge, MA: Harvard University Press, 1988.

Mayr, Ernst, and William B. Provine, eds. *The Evolutionary Synthesis: Perspectives on the Unification of Biology*. Cambridge, MA: Harvard University Press, 1998.

McMenamin, Mark A. S. "The Origins and Radiation of the Early Metazoa." In Allen and Briggs, *Evolution and the Fossil Record.*

Mendelson, Tamra C., and Kerry L. Shaw. "Rapid Speciation in an Arthropod." *Nature* 433 (January 2005): 375–76.

Mensh, Elaine, and Harry Mensh. *The IQ Mythology: Class, Race, Gender, and Inequality*. Carbondale: Southern Illinois University, 1991.

Meyer, Stephen C. "The Origin of Biological Information and the Higher Taxonomic Categories." In Dembski, *Darwin's Nemesis*, pp. 174–213.

Midgley, Mary. *Evolution as a Religion: Strange Hopes and Stranger Fears.* London: Methuen, 1985.

Miller, Carolyn R., and S. Michael Halloran. "Reading Darwin, Reading Nature; Or, On the Ethos of Historical Science." In Selzer, *Understanding Scientific Prose*, pp. 106–26.

Mohan, Matthew and Bernard Linsky. "Introduction." In Mohan and Linsky, eds., *Philosophy and Biology*. Calgary: University of Calgary, 1988.

Monod, Jacques. *Chance and Necessity: An Essay on the Natural Philosophy of Modern Biology*. New York: Vintage, 1972.

Montagu, Ashley, ed. *Race and IQ*, expanded ed. New York: Oxford University Press, 1999.

Moore, John A. *Science as a Way of Knowing: The Foundations of Modern Biology*. Cambridge, MA: Harvard University Press, 1993.

Morris, Richard. *The Evolutionists: The Struggle for Darwin's Soul*. New York: W. H. Freeman, 2001.

Morris, Simon Conway. *The Crucible of Creation: The Burgess Shale and the Rise of Animals*. Oxford: Oxford University, 1998.

———. *Life's Solution: Inevitable Humans in a Lonely Universe*. Cambridge: Cambridge University Press, 2003.

Nelson, Paul A. "The Rose of Theology in Current Evolutionary Reasoning." In Pennock, *Intelligent Design Creationism*, pp. 677–704.

Neumann-Held, Eva M., and Christoph Rehmann-Sutter, eds. *Genes in Development: Re-Reading The Molecular Paradigm*. Durham, NC: Duke University Press, 2006.

Olson, Steve. *Mapping Human History: Genes, Race, and Our Common Origins*. Boston: Houghton Mifflin, 2002.

Orr, H. Allen. "The Population Genetics of Speciation: The Evolution of Hybrid Incompatibilities." *Genetics* 139 (April 1995): 1805–1813.

Page, Benjamin I. *Choices and Echoes in Presidential Elections: Rational Man and Electoral Democracy*. Chicago: University of Chicago Press, 1978.

Palevitz, Barry A. "Science Versus Religion: A Conversation with My Students." In Kurtz, *Science and Religion*, pp. 171–79.

Paul, Christopher R. C. "Patterns of Evolution and Extinction in Invertebrates." In Allen and Briggs, *Evolution and the Fossil Record*, pp. 99–121.

Pennock, Robert T., ed. *Intelligent Design Creationism and Its Critics: Philosophical, Theological, and Scientific Perspectives*. Cambridge, MA: MIT Press, 2001.

Pinker, Steven. *The Blank Slate: The Modern Denial of Human Nature*. New York: Viking, 2002.

Plantinga, Alvin. "When Faith and Reason Clash: Evolution and the Bible." In Pennock, *Intelligent Design Creationism*.

Poppe, Kenneth. *Reclaiming Science from Darwinism: A Clear Understanding of Creation, Evolution, and Intelligent Design*. Eugene, OR: Harvest House, 2006.

Popper, Karl. *Conjectures and Refutations: The Growth of Scientific Knowledge*. New York: Harper and Row, 1965.

Prindle, David F. *The Paradox of Democratic Capitalism: Politics and Economics in American Thought*. Baltimore: Johns Hopkins University Press, 2006.

Proudhon, Pierre Joseph. "What Is Property?" In Albert Fried and Ronald Sanders, eds., *Socialist Thought: A Documentary History*. Garden City, NY: Doubleday, 1964, pp. 201–203.

Pugliucci, Massimo. *Denying Evolution: Creationism, Scientism, and the Nature of Science*. Sunderland, MA: Sinauer Associates, 2002.

Putnam, Hillary. *The Collapse of the Fact/Value Dichotomy and Other Essays*. Cambridge, MA: Harvard University Press, 2002.

Queller, David C. "The Spandrels of St. Marx and the Panglossian Paradox: A Critique of a Rhetorical Programme." *Quarterly Review of Biology* 70, no. 4 (December 1995): 485–89.

Rainger, Ronald. *An Agenda for Antiquity: Henry Fairfield Osborn and Vertebrate Paleontology at the American Museum of Natural History, 1890–1935.* Tuscaloosa: University of Alabama, 1991.

Rana, Fazale, and Hugh Ross. *Origins of Life: Biblical and Evolutionary Models Face off.* Colorado Springs, CO: NavPress, 2004.

Rawls, John. *A Theory of Justice*. Cambridge, MA: Harvard University Press, 1971.

Reeve, H. Kern, and Laurent Keller. "Levels of Selection: Burying the Units-of-Selection Debate and Unearthing the Crucial New Issues." In Laurent Keller, ed., *Levels of Selection in Evolution.* Princeton, NJ: Princeton University Press, 1999, pp. 3–14.

Regal, Brian. *Henry Fairfield Osborn, Race, and the Search for the Origins of Man.* Burlington, VT: Ashgate, 2002.

Richerson, Peter J., and Robert Boyd. *Not by Genes Alone: How Culture Transformed Human Evolution*. Chicago: University of Chicago Press, 2005.

Roe v. Wade, 410 U. S. 113 (1973).

Rosenberg, Alex. *The Structure of Biological Science*. Cambridge: Cambridge University, 1985.

Ruse, Michael. *Darwin and Design: Does Evolution Have a Purpose?* Cambridge, MA: Harvard University Press, 2003.

———. "Evolutionary Ethics: Healthy Prospect or Lost Infirmity?" In Mohan and Linsky, *Philosophy and Biology*, pp. 27–73.

———. "Is the Theory of Punctuated Equilibria a New Paradigm?" In Somit and Peterson, *The Dynamics of Evolution.*

———. "Methodological Naturalism Under Attack." In Pennock, *Intelligent Design Creationism*, pp. 364–85.

———. *Mystery of Mysteries: Is Evolution a Social Construction?* Cambridge, MA: Harvard University Press, 1999.

———. *Taking Darwin Seriously: A Naturalistic Approach to Philosophy*, 2nd ed. Amherst, NY: Prometheus, 1998.

Rushton, J. Philippe. *Race, Evolution, and Behavior: A Life History Perspective.* New Brunswick, NJ: Transaction, 1995.

———. "Race, Intelligence, and the Brain: The Errors and Omissions of the "Revised" Edition of S. J. Gould's *The Mismeasure of Man.*" *Personality and Individual Differences* 23, no. 1 (1997): 169–80.

Rushton, J. Philippe, and Arthur R. Jensen. "Commentary: The Totality of Available Evidence Shows the Race-IQ Gap Still Remains." *Psychological Science* 7, no. 10 (October 2006): 921–22.

———. "Thirty Years of Research on Race and Cognitive Ability." *Psychology, Public Policy, and the Law* 11, no. 2 (2005): 235–64.

Saad, Gad. "Evolution and Political Marketing." In Somit and Peterson, *Human Nature and Public Policy*, pp. 121–38.

Sabine, George H. *A History of Political Theory*. New York: Henry Holt, 1950.

Santa Fe Independent School District v. Doe, 120 S. Ct. 2266 (2000).

Schattschneider, E. E. *The Semi-Sovereign People*. New York: Holt, Rinehart and Winston, 1960.

Schellenberg, E. Glenn. "Music Lessons Enhance IQ." *Psychological Science* 15, no. 8, (August 2004): 511–14.

Scott, Eugenie C. *Evolution vs. Creationism: An Introduction*. Westport, CT: Greenwood Press, 2004.

———. "The 'Science and Religion Movement': An Opportunity for Improved Public Understanding of Science?" In Kurtz, *Science and Religion*, pp. 111–25.

Searle, John R. *The Construction of Social Reality*. New York: Simon and Schuster, 1995.

Segerstrale, Ullica. *Defenders of the Truth: The Sociobiology Debate*. Oxford: Oxford University Press, 2000.

Seligman, Daniel. *A Question of Intelligence: The IQ Debate in America*. New York: Birch Lane, 1992.

Selzer, Jack, ed. *Understanding Scientific Prose*. Madison: University of Wisconsin Press, 1993.

Shanahan, Timothy. *The Evolution of Darwinism: Selection, Adaptation, and Progress in Evolutionary Biology*. Cambridge: Cambridge University Press, 2004.

Shermer, Michael. *Why Darwin Matters: The Case against Intelligent Design*. New York: Henry Holt, 2006.

Shipman, Pat. *Taking Wing: Archaeopteryx and the Evolution of Bird Flight*. London: Phoenix, 1998.

Simon, Julian L., and Paul Burstein. *Basic Research Methods in Social Science*, 3rd. ed. New York: Random House, 1985.

Simpson, George Gaylord. *The Major Features of Evolution*. New York: Columbia University Press, 1953.

———. *Tempo and Mode in Evolution*. New York: Columbia University. 1944.

Singer, Peter. *A Darwinian Left: Politics, Evolution, and Cooperation*. New Haven, CT: Yale University Press, 2000.

———. *This View of Life: The World of an Evolutionist*. New York: Harcourt, Brace, and World, 1964.

Smith, John Maynard, and Eors Szathmary. *The Origins of Life: From the Birth of Life to the Origin of Language*. Oxford: Oxford University Press, 1990.

Snyderman, Mark, and Stanley Rothman. *The IQ Controversy: The Media and Public Policy*. New Brunswick, NJ: Transaction, 1988.

Sober, Elliott. *The Nature of Selection*. Cambridge, MA: MIT Press, 1984.

Sober, Elliott, and Richard Lewontin. "Artifact, Cause, and Genic Selection." In Boyd, Gasper, and Trout, *Philosophy of Science*, pp. 571–88.

Somit, Albert, and Steven Peterson. *Darwinism, Dominance, and Democracy: The Biological Bases of Authoritarianism*. Westport, CT: Praeger, 1997.

———. *The Dynamics of Evolution: The Punctuated Equilibrium Debate in the Natural and Social Sciences.* Ithaca, NY: Cornell University Press, 1992.

———. *Human Nature and Public Policy: An Evolutionary Approach.* New York: Macmillan, 2003.

Standish, Timothy G. "Cutting Both Ways: The Challenge Posed by Intelligent Design to Traditional Christian Education." In Dembski, *Darwin's Nemesis*, pp. 117–34.

Stanley, Steven. "Macroevolution and the Fossil Record." *Evolution* 36, no. 3 (1982): 460–73.

———. *Macroevolution: Pattern and Processes*. San Francisco: W. H. Freeman, 1979.

Stebbins, G. Ledyard, and Francisco J. Ayala. "Is a New Evolutionary Synthesis Necessary?" *Science* 213, no. 4511 (August 28, 1981): 967–71.

Steele, Edward J., Robyn A. Lindley, and Robert V. Blanden. *Lamarck's Signature: How Retrogenes Are Changing Darwin's Natural Selection Paradigm*. Reading, MA: Perseus, 1998.

Stenhouse, David. *The Evolution of Intelligence: A General Theory and Some of Its Implications*. London: George Allyn and Unwin, 1973.

Sterelny, Kim. *Dawkins vs. Gould: Survival of the Fittest*. Cambridge: Totem Books, 2001.

Sternberg, Robert J., and Douglas K. Detterman, eds. *What Is Intelligence? Contemporary Viewpoints on Its Nature and Definition.* Norwood, NJ: Ablex, 1986.

Stone v. Graham, 449 U. S. 39 (1980).

Strobel, Lee. *The Case for a Creator: A Journalist Investigates Scientific Evidence That Points toward God.* Grand Rapids, MI: Zondervan, 2004.

Sundquist, James L. *Dynamics of the Party System Alignment and Dealignment of Political Parties in the United States.* Washington, DC: Brookings Institution, 1983.

Thompson, D'Arcy. *On Growth and Form*. Cambridge: Cambridge University Press, 2004.

Thorndike, Robert M., with David F. Lohman. *A Century of Ability Testing*. Chicago: Riverdale, 1990.

Thurow, Lester C. *The Future of Capitalism: How Today's Economic Forces Shape Tomorrow's World.* New York: Penguin, 1996.

Tingley, Dustin. "Evolving Political Science: Biological Adaptation, Rational Action, and Symbolism." *Politics and the Life Sciences* 25, no. 1–2 (January 2007): 23–40.

Tooby, John, and Leda Cosmides. "The Psychological Foundations of Culture." In Barkow, Cosmides, and Tooby, *The Adapted Mind*, pp. 19–136.

Toulmin, Stephen. *The Uses of Argument*. Cambridge: Cambridge University Press, 1958.

Van Till, Howard J. "When Faith and Reason Cooperate." In Pennock, *Intelligent Design Creationism*, pp. 147–63.

Vernon, Philip E. *Intelligence: Heredity and Environment*. San Francisco: W. H. Freeman, 1979.

Vrba, Elisabeth S., and Niles Eldredge. "Individuals, Hierarchies and Processes: Towards a More Complete Evolutionary Theory." *Paleobiology* 10, no. 2 (1984): 146–71.

Waxman, D., and S. Gavrilets. "20 Questions on Adaptive Dynamics." *Journal of Evolutionary Biology* 18 (2005): 1139–54.

Weinberg, Steven. *Facing Up: Science and Its Cultural Adversaries*. Cambridge, MA: Harvard University Press, 2001.

Weiner, Jonathan. *The Beak of the Finch*. New York: Random House, 1995.

Weis, Susanne, and Heinz-Martin Sub. "Reviving the Search for Social Intelligence—A Multitrait-Multimethod Study of Its Structure and Construct Validity." *Personality and Individual Differences* 42, no. 1 (January 2007): 3–14.

Wells, Jonathan. "Common Ancestry on Trial." In Dembski, *Darwin's Nemesis*, p. 171.

Whitcomb, John C., and Henry M. Morris. *The Genesis Flood: The Biblical Record and Its Scientific Implications*. Phillipsburg, NJ: Presbyterian and Reformed Publishing Company, 1961.

White, Elliott, ed. *Sociobiology and Human Politics*. Lexington, MA: D. C. Heath, 1981.

Wilson, David Sloan. *Darwin's Cathedral: Evolution, Religion, and the Nature of Society*. Chicago: University of Chicago Press, 2002.

Wilson, Edward O. "Apocalypse Now: A Scientist's Plea for Christian Environmentalism." *New Republic*, September 2006, pp. 4, 18.

———. *Consilience: The Unity of Knowledge*. New York: Random House, 1999.

———. *The Diversity of Life*. Cambridge, MA: Harvard University Press, 1992.

———. *On Human Nature*. Cambridge, MA: Harvard University Press, 2004.

———. *Naturalist*. Washington, DC: Warner Books, 1994.

———. *Sociobiology: The New Synthesis*, 25th anniversary ed. Cambridge, MA: Harvard University Press, 2000.

Wright, Robert. *The Moral Animal: The New Science of Evolutionary Psychology.* New York: Pantheon, 1994.

———. *Nonzero: The Logic of Human Destiny.* New York: Random House, 2000.

Wright, Sewell. "Character Change, Speciation, and the Higher Taxa." *Evolution* 36 (1982): 427–43.

Zaller, John R. *The Nature and Origin of Mass Opinion.* Cambridge: Cambridge University Press, 1992.

Index